History of Technology

History of Technology

Third Annual Volume, 1978

Edited by

A. RUPERT HALL and NORMAN SMITH

Imperial College, London

MANSELL
London 1978

ISBN 0 7201 0813 6
ISSN 0307-5451

Mansell Information/Publishing Limited, 3 Bloomsbury Place,
London WC1A 2QA

First published 1978

© Mansell and the Contributors, 1978

All rights reserved.
No part of this publication may be reproduced or transmitted in any form or by any means, electronic or mechanical, including photocopy, recording, or any information storage and retrieval system, without permission in writing from the publishers.

Articles appearing in this publication are abstracted and indexed in *Historical Abstracts* and *America: History and Life*.

British Library Cataloguing in Publication Data
History of technology.
   3rd annual vol. : 1978.
   1. Technology — History — Addresses, essays, lectures
   I. Title II. Hall, Alfred Rupert
   III. Smith, Norman, b.1938
   609   T15

ISBN 0-7201-0813-6

Typeset by
Preface Ltd., Salisbury, Wiltshire
Printed in Great Britain by
The Scolar Press, Ilkley, West Yorkshire

Contents

Preface	vii
JACK SIMMONS **Technology in History**	1
R. A. BUCHANAN **History of Technology in the Teaching of History**	13
P.B. MORICE **The Role of History in a Civil Engineering Course**	29
JOYCE BROWN **Sir Proby Cautley (1802-1871), a Pioneer of Indian Irrigation**	35
A. RUPERT HALL **On Knowing, and Knowing how to ...**	91
FRANK D. PRAGER **Vitruvius and the Elevated Aqueducts**	105
JAMES A. RUFFNER **Two Problems in Fuel Technology**	123
JOHN C. SCOTT **The Historical Development of Theories of Wave-Calming using Oil**	163
The Contributors	187

Preface

It is hard to realize that the time has come, for the third time, to offer our readers a collection of papers. The first three, by Messrs. Simmons, Buchanan and Morice, were read at a meeting on the teaching of the history of technology held at Imperial College on 19 March 1977. The remaining papers consider a variety of rather unusual themes ranging from Roman aqueducts to the problem of urban-railway smoke-pollution.

We are pleased to welcome two contributions from the United States of America in this issue, but sadly we have to report the death of Frank D. Prager in June 1978, just before the volume went to press. As before, correspondence concerning publication in this series may be addressed to the editors at the Department of History of Science and Technology, Sherfield Building, Imperial College, London SW7 2AZ.

A. RUPERT HALL
NORMAN A. F. SMITH

Technology in History

JACK SIMMONS

My purpose is to say something about the part that the study of technology can play in the study of history as a whole. How far are technological explanations useful, relevant, necessary, even essential to an understanding of the way things went in the past, and of the lives that people lived? Here at once let me make a distinction. It must I think be perfectly clear that to understand *lives*, the ordinary activities of human beings in ages other than our own, it is indispensable to consider the technologies that served them, for they formed in many respects the very framework of those lives themselves. They may have been, to our way of thinking, very simple technologies. But the simplicity is relative to experience, to what went before: the advance to the wheel, whether on the ground or in the potter's hands, cannot be considered a less striking achievement than the development of the telephone or the radio in the past hundred years. The changes that new technologies wrought in the lives of those who could make use of them are evidently great, perhaps here and there fundamental. If we wish to see those people in the past, remote or near, clearly and fully we must investigate the apparatus with which their lives were lived: surely one of the things that suggested itself in looking at what survives of Pompeii.

I am not going to say much on this theme in this paper, for I take it we shall all understand its importance. I want rather to look at something a little less obvious, certainly something that has been less attended to: namely the importance of technological development and change in some broad historical processes. For I think these things have often been forgotten, or insufficiently remembered, at many points ignored or misunderstood. Although I shall range fairly widely here and there, it may help if I keep to a single central theme: the subject of communications, using that word in its broadest sense to mean not transport alone but all forms of communication down to the telecommunications of the modern world.

Anyone who has studied political history, the history of government and states, or tried to teach it, needs to recognize at the outset and never forget one axiom: that what really happened was not necessarily the same thing as was intended to happen, that laws might be passed and be imperfectly obeyed or even be dead from the moment they went into force. The gulf between theory and practice is often enormous. It is clear beyond any argument, for example, that certain Acts were passed in London between 1530 and 1560 ordaining changes in religious observance: the destruction of roods and rood screens and of pictures in churches and so on, then their restoration under Mary, and their destruction again under Elizabeth. But how far were these measures enforced? What are we to make of the survivals, still to be seen now, of some of these things in the remoter parts of the country, on the Welsh border and in the North? It is perfectly plain that the government in London could not ensure that its decrees were obeyed; it had no means of knowing, directly or immediately, that they had been carried out. It was the better part of a week's journey from London to the wilder parts of Herefordshire — a fortnight there and back, allowing a little time for investigation on the spot. Think of what that meant, in terms of the quality of government: the inability — so far from what we now take for granted — to know, by any instant means, that the order has arrived and is being obeyed.

The passing of laws fills the books concerned with political history. It has its own interest as a reflection of policy and intention, of thinking. It may be far less interesting than the consideration of the ways in which they were or were not implemented. The first thing needed is a firm grasp of the means of communication at the disposal of those whose job it was to see that the laws were carried out. For anyone who sets himself to study the history of a state, of France or Russia or the emergent United States in the nineteenth century, it ought to be an immediate task to discover how long it took to communicate between the capital and the different parts of the country and what the means of communication were. Were they, for instance, liable to interruption in the winter?

Let me pursue that last question for a moment, with one case in mind. The British colonies in North America, the nucleus of what we call Canada today, were politically troubled in the 1830s; troubled in their relations with one another and with the distant government in London. A minor rebellion was the result in 1837, and in the following year Lord Durham was dispatched to investigate what was wrong: an able, awkward man who asked searching questions and did not mind if the answers he found to them were disagreeable. When he set out, he had in mind one main political solution: a federal union of all the five separate colonies. It was an intelligent idea that suggested itself naturally and was eventually realized. But when Durham looked into things on the spot he changed his mind and rejected it. One of the chief obstacles to any such plan was that the St Lawrence was frozen over for half the year, which cut off the colonies near its mouth — Prince Edward Island, New Brunswick, and Nova Scotia — from the rest, from the old French colony of Canada and its western offshoot in what we now call Ontario. As early as 1838 Durham perceived that a political union would not work satisfactorily until a railway had been built linking the colonies together — a railway that could be kept open all through the year and provide the continuous communication that any kind of united government required.[1] This idea showed real prescience, the imagination of a truly forward-looking mind, for when Durham left

England the first trunk railway, the Grand Junction, had been open less than a year and the London & Birmingham was six months from completion. But he saw and said in the famous report published in 1839 that the construction of a railway was an essential element in a most desirable political change. He was right. The railway was built slowly and painfully over the next forty years. The federal idea came up again in the 1850s. With railways actually under construction west of Montreal and a bridge being built across the Saint Lawrence there, it became realistic to talk once again about union. After much discussion and some wrangling the federation was achieved in 1867. The last section of the railway was completed nine years later. Here we see politics influenced and in part determined by technology, and here is a politician recognizing that a technology only just emerging will be a powerful force in a development he advocates for the future.

Let us now look at another kind of government, one of vast scale stretching over half the world, one that was profoundly affected — certainly at one point radically changed — by technological development. I mean British rule in India: by the government and the East India Company in partnership from 1784 to 1858, and then by the government alone. In the late eighteenth and the early nineteenth centuries, could it really be said that British India was governed from Britain at all? By ships sailing round the Cape of Good Hope — the East Indiamen, majestic and slow — the voyage lasted at least four months and often six. It regularly took a year or more to receive the reply to a letter sent to India from London. In those conditions no order, however peremptory, could be enforced. It was inevitable, in the nature of the case, that the Governor-General and his colleagues at Calcutta should decide what ought to be done in the shifting conditions of Indian politics and justify it to the home authorities afterwards. If they disagreed they could only say so in another slow exchange of correspondence, or in an extreme case like that of Lord Ellenborough in 1844, recall the offender to England and dismiss him. But that was a retrospective punishment.

By the time of the Ellenborough incident, Britain and India were being brought closer in time through new developments in transport. The Overland Route, by way of the Mediterranean, was already in use: P. & O. steamship from Southampton to Alexandria, across the Isthmus of Suez, then by steamer again down the Red Sea, or of course after 1869 through the Suez Canal. This reduced the time taken on the journey to a minimum of about six weeks. But even so, three months had to go by at least before an answer could be received to a letter written in London; the separation of real power between the home government and its representatives in India was still dangerously great. How dangerously became clear in 1857 when the news of the Mutiny arrived home, and the very continuance of the British Raj in India was at risk until military reinforcements could get there. The East India Company was wound up in the following year, and its power transferred wholly to the government; but that did nothing to solve the problem of communication.

The way forward was already being shown. Submarine cables from Britain to France and Ireland were brought into use successfully in 1852-3. A company was registered to carry a cable under the Atlantic to America in 1856 and in the very year of the Indian Mutiny work on it began. It took nine years to complete the task, years that included the American Civil War. When the triumph was complete in 1866 it initiated the rapid development of a network crossing the world. The cable reached Bombay in 1870, and it brought a political revolution in its train. Now at last the government in London could really control the government of India. Quite soon a great change in the whole style of the régime there came to be noticed. The Governor-General, though he could still be in ordinary times one of the most powerful men in the world, was clearly subject to the decisions of the British Cabinet. In one field after another it became plain that the grip of Britain on India was being tightened: in frontier policy, and in economic affairs where the interests of the cotton industry seemed to be more and more subordinated to those of Lancashire. Before 1870 such policies could have been laid down on paper, yet not implemented. In 1870 direct rule had become possible.

We must not pursue the consequences further. They stretch very far. The nationalist movement of the 1880s must be seen as, in part, a protest against an increasingly rigid foreign control. But it was not any decree of government, it was the capitalists, engineers and servants of the cable company whose labours had brought that about.

The political and economic changes wrought by telecommunications have been profound — from the early developments of the electric telegraph in Britain, sensationally floodlit in 1845 by the arrest of a murderer at Paddington station.[2] There is a whole literature, of course, dealing with the techniques involved and their successive development. We have one useful study concerned largely with the attitude of the state towards the telegraph; its purchase of the private companies from 1868 onwards may be taken as the first measure of nationalization in the form that is familar to us now in this country.[3] But nobody has thought to write a comprehensive study of the effects of this great development, of telecommunication in history.

Here, as always, we must be careful not to exaggerate. Even with the aid of the telegraph and the telephone and their still more sophisticated successors, governments are not always in control of the people and the forces they pretend to guide, that is only too evident. The development of the means of control does not ensure that it shall be used effectively. It can indeed be argued, and I think very cogently, that the multiplication and improvement of the technologies placed in the hands of politicians, administrators, and business men have made their task harder by increasing the range of choice in front of them. Although that is not the theme of this paper, it needs to be mentioned in passing. All I am trying to do is to indicate a few points at which technological development opened up new opportunities to men in power, enlarged that range of choice; and I want now, pursuing this further, to look at three other cases of different sorts. The first involves economic power. In the second not one but a succession of these developments played a considerable part in the relations between three states and eventually between one of them especially — Great Britain — and the rest of the world. The third is a chapter both of political and of medical history.

In Britain we are familiar with the troubles that beset agriculture in the 1870s, and we attribute them largely to the export of grain from the North American continent and from Russia, which flooded Europe and damaged Britain particularly because she refused to reimpose tariffs to shut out the wheat from abroad. The threat materialized just then, as we can see, through the opening up of central and western Canada and of the great plains of Russia by the building of railways. But what about the United States? Railways had criss-crossed the grain-growing States of Indiana and Illinois in the 1850s, and they were already stretching out fast into Iowa and Wisconsin. Why did this export not reach its 'flood' for another twenty years? In large measure because the railway system, which looked continuous on paper, was interrupted by numberless breaks of gauge; the bulk of the grain still moved eastwards slowly by water. But water transport could not handle all the traffic, even with the assistance of the railways — not more than two-thirds of the traffic, at a maximum, at the end of the Civil War. A revolution followed, which was purely technological in character. The railway system of the West and North was physically unified by the removal of breaks of gauge; and then it became possible to move an ever-increasing quantity of grain quickly and cheaply to the eastern seaports. The American invasion of the markets of Europe could begin, now and not before; and this technological advance was the reason.[4]

My second case is of a different sort. The naval supremacy that Britain forged for herself in the Napoleonic wars did not arise in any large measure from superior skill in the techniques of building or equipping her ships; the superiority appeared in the handling of them. Before those wars ended steam propulsion was being applied to commercial ships with success. By 1830 steamships were a familiar sight on the Thames and in the English Channel. But the new power was extended to warships very tentatively. Marc Brunel persuaded the Admiralty to build a steam warship in 1822, but it and its early successors were regarded in the Navy with widespread distrust and contempt. Special difficulties presented themselves, it must be allowed: all steamships were extravagantly expensive in service until the development of the compound engine in the 1860s drastically reduced the deadweight of fuel they had to carry in order to operate over long distances. Moreover, the paddle steamer was clearly vulnerable in war, its power concentrated at a single point that provided a large and easy target to other ships' guns.

When the politicians and naval administrators looked into these developments they acted quite rapidly on one, but not on the other. The adoption of screw propulsion followed fast on the first experiments, which were made in 1837. In 1843 — the year in which I.K. Brunel's *Great Britain* made its memorable first crossing of the Atlantic — the Admiralty commissioned H.M.S. *Rattler*, the first naval vessel in Europe to be propelled by a screw. Successive improvements followed over the next five years, in the warships *Amphion* and *Ajax*, and thereafter the British Admiralty abandoned the paddle-steamer altogether. Here was a quick apprehension of a technological change, promptly acted on.

But, in warships as in the merchant navy, the transition from sail to steam was much less rapid than we are apt to suppose, for which there were a number of explanations. The capital cost of abandoning the whole British fleet and turning to new ships was too great for the politicans to contemplate. The steamship was expensive and cumbrous and it required a chain of reliable coaling stations across the world, which was achieved gradually; hence, for example, the government's willingness to accept control of Aden in 1839. (Its very origins as a British colony are, or were until very lately, commemorated by the name of the headland there, Steamer Point.) From henceforth this was a distinct thread in the colonial policy of Britain throughout the nineteenth century. Successive governments desired to avoid the expansion of the Empire, or to keep it down to the minimum. They found themselves forced nevertheless to increase their responsibilities, and one reason lies in the voracious consumption of coal by the steamships they had to provide for — their own, in the Navy, and those of British traders who looked to them for assistance. From this point of view the economies in the consumption of fuel effected by triple and quadruple expansion from 1874 onwards were of political, as well as economic and technical importance.

In the history of warships there is a direct and clear line of development over the seventy years from 1843 to the First World War. Here are a few of its most conspicuous landmarks: the adoption of armour plating around 1860, followed quickly by a great increase in the firing power of naval guns, leading to a 'race between ordnance and armour plate';[5] the change-over from iron to steel, heralded by blockade-runners in the American Civil War and accepted by the British Admiralty in 1877; the grouping of guns in a turret and the development of the torpedo, both again taking their origins in the American war and applied in Europe subsequently; the triumphant appearance of the steam turbine in 1895, installed almost at once in British torpedo boats; and the final combination of all these changes in H.M.S. *Dreadnought* in 1906.

In nearly every one of these instances, if the new development was not British in origin, it was taken up and made effective as a force in the naval policies of the world by the British Admiralty. But at what cost! — a very obvious cost in one sense, a concealed and even more important one in another. In 1852 the country's expenditure on the Navy was £5 million. In the years 1860-84 it stood fairly steadily at about twice that amount. Then began a staggering rise. The cost of the Navy in 1913 was over four times what it had been thirty years before. Or, to look at it differently, whereas in 1883 an eighth of the entire revenue of the state was spent on the Navy, by 1913 that proportion had almost doubled, to twenty-four per cent.[6]

Not all this increase is attributable to new, more elaborate and costly equipment; much of it arose from the mere enlargement of the Navy to meet fresh dangers, above all after 1898 when the fatal race with Germany began. But never at any time before I think had technical changes forced so great an increase of expenditure on a government to be incurred so quickly.

The hidden consequence was subtler, and to those few politicians who understood it more alarming still. Though Britain, with some set-backs and stupidites and follies, managed to keep as a rule the lead in this fearful rivalry — superbly demonstrated in *Turbinia* and *Dreadnought* — she was in a sense exposing herself to new dangers each time she committed herself to a new technique or device. The French seized the initiative for a moment in 1859 with their battleship *La Gloire*, with an armour-plated wooden hull. Britain went one better in 1860 with H.M.S. *Warrior*, in which the entire hull was of iron. But the French, or any other power, could play at that game too if they chose. Each time one of these decisive advances occurred even if Britain pioneered them, her whole existing navy was placed at risk; and as it was much the largest navy in the world and she was more dependent on it for her power than any other state, her risk was greater. When it came to the most critical stage of all in the fifteen years preceding the first war with Germany, the Admiralty made relentless demands on the Treasury, and they became fiercest of all when the Germans began to build *Dreadnoughts* of their own: for ships of that kind had made all other extent battleships obsolete. Here is one of the vital elements in the Parliamentary crisis of 1909. Lloyd George did not frame his famous budget of that year simply in order to provoke the House of Lords into attacking it. He began, as Chancellor of the Exchequer, with the need to find vast new sums for more *Dreadnoughts* (summed up in the jingoist cry, 'We want eight and we won't wait'), not to mention submarines and other equipment of new types. In the three years 1909-12 alone Britain's naval expenditure leapt up by a third. And in the end, when the test came, the margin of safety was very narrow indeed. The Germans might well have won the Battle of Jutland in 1916; in 1917 their submarines reduced Britain, as the Cabinet knew, to a prospect of starvation within six weeks.

And now for the last of these cases — quite unlike the other two. It takes us to West Africa. From the fifteenth century to the nineteenth Europeans had always found it difficult, often impossible, to work there on account of the climate and the diseases that attacked them. In spite of that the British Government persisted in its determination to destroy the slave trade on that coast (at a very considerable price in British lives in the end, if that could ever be reckoned up). This resolution involved two actions which were complementary. The first was the policing of the coast, based here and there on political control. The second was the penetration of the interior, towards the sources from which the slaves were drawn, with a view to developing gradually new trades to take the place of the one that was to be destroyed. The first of these objectives was proclaimed when the British Government entered on a protectorate over Lagos in 1851, which became complete control ten years later. That was a political decision, and it is in all the textbooks. But it was hardly, in the long run, more important than something else, which took place at the same time and figures, so far as I know, in no textbooks at all. An exploring voyage was made up the River Niger in a small steamship, *Pleiad*, in 1854-6, under the command of a young Orkney surgeon, W.B. Baikie. It penetrated 250 miles further up the river than any Europeans had done before and accumulated much useful knowledge of the country. One piece of knowledge was more valuable than all the rest. Baikie insisted firmly that every man on board should take regular doses of quinine, as a prophylactic against malaria, and not one of them died of the disease during the whole expedition. Baikie had proved that it was possible for Europeans to survive even on pestilential West African rivers.[7] He went on to demonstrate that possibility again on a second expedition. Baikie had in fact gone a long way towards solving one of the most crucial of all the problems in the penetration of the Tropics. Not the whole way — much more was to be done by Manson and Ross, and by other workers in the next generation — but a long way. Baikie was a quiet man without the slightest touch of bombast or self-advertisement. Perhaps in a way, *too* quiet, for his achievement passed with very little recognition at the time and the lessons to be drawn from it were not sufficiently learnt. But looking back now we may well feel that his application of a technique in combating disease was as significant as the annexation of Lagos, the political act that stands out in history, conventionally told.

These then are a few cases — a few out of very many that could be adduced, confined to the nineteenth and twentieth centuries and almost wholly to the field of communications — cases in which, it seems to me, technological explanations are required by the intelligent student of history if he is to get his study right. I am not of course suggesting that these explanations have never been given before, or that my points are in any way original. I *am* suggesting that they have not, as a rule, received the prominence or the clarity of treatment that they deserve, and that (especially in the example I drew from naval practice) the clear chain of causes and consequences has not been satisfactorily shown as such by some of the historians who are concerned to explain national policies.

I also think that many students of technology have regarded their piece of technology as a study on its own, have failed to see it in its true historical context, and so sometimes have missed part of what is most interesting and profoundly important in the technology itself. I speak from the historians' side of the fence; I have neither training nore expertise in any kind of technology. But I have long been aware of the need to search for explanations, in engineering or in some other branch of science, where there is a chance that they may help me; and I am sure that those explanations have often, if I have understood them, enabled me to grasp more adequately the political or economic or social matters I was examining.

I suggest the key to much profitable exploration is to persist in asking one question, not merely 'What happened?' nor 'How was it made to happen?', but 'What were its effects?' It constantly surprises me that people are so incurious in that matter. They are perhaps carried away by the sudden appearance of a new machine, a new device, a new invention as it is commonly called. They fall to asking 'Who produced this, and how did he produce it?' — natural questions and good ones, but not the only ones that need asking. Sometimes, really, not the questions that matter most, but less important, if you think about it, than discovering the differences that the new development made in appearance or reality. I look at machines in museums. I am often told a great deal there, and it may be fascinating, about the process of thought and experiment that produced them. But I am seldom informed of the changes that they wrought: what they cost to build and run as compared with their predecessors, their productivity, the effects they had on employment, the side-effects they sometimes entailed in local or national politics, in the relations between states. Without the consideration of those effects, the whole exposition is diminished; the machine one is looking at becomes a curiosity, an antiquity, no more. It lacks a place in history, a place in life.

The mere fact that some technique or some machine was the first of its kind may not of itself mean very much; or rather, what it means changes according to one's point of view. We shall always be tantalized by knowing so little about Trevithick's Penydarren locomotive, for it represented a feat of original genius and its fame reached out beyond South Wales. But one of the things we do know about it is that it broke the light track it ran on and was abandoned in consequence. It represented a brilliant idea, imperfectly executed (presupposing, among other things, the further development of the iron rail) and it came to a dead stop. The real breakthrough to the locomotive able to run smoothly on its track at high speeds came with *Rocket* and its immediate successors twenty-five years later. Trevithick's thinking was taken up and developed elsewhere, as the basis for further experiment; the full realization of the idea, the turning of it into effect, is the province of the Stephensons. It is in 1829-30 that the locomotive is seen to have become efficient, versatile, dependable, still capable of much further improvement, indeed, but now ready for work. When all has been said in justice to the early pioneers, this is the point of take-off — the point at which investors and commercial men, even governments, begin to feel they must take the railway seriously. It is so because it is the point at which the locomotive and the railway begin to have evident, discoverable, and far-reaching effects. Those are the moments in which the student of history is bound to be most interested, whether he is a student of technology or of economic or political life, the significant moments when development goes forward in a new direction.

To pursue this subject a little further for a moment: if I am shown a locomotive in a museum, at York or Hamar or Nuremberg or Baltimore, I want to learn a number of things about it. I want of course a technical description; and if the machine was demonstrably bigger or faster or in some significant way more sophisticated, as a machine, than its predecessors, I want to be told that too. I then want an answer to the question: why is this machine here? What is its place in the complex of evolution of the locomotive, in Britain or Europe or the world? It may of course have come to be in the museum almost by accident, as the gift of some kind benefactor or an example of local industry, not in itself of any wider significance. Or it may be there as a really important machine in its own right, demonstrating a stage of technological development that cannot now be seen anywhere else in the round: like *Lion* at Liverpool or *Coppernob* at York or the Metropolitan Railway steam locomotive at Syon Park. Or again it may be there (and this can very well justify its preservation) not as any kind of pioneer but for the opposite reason: because it was characteristic of its time, not exceptional — the temptation is always to cherish the rarity, anything curious, and to neglect what is typical, what ordinary people commonly knew and used. These things particularly need to be watched for and kept: the Dean Goods engine at Swindon, the South Wales coal tank engine now at Caerphilly.

Here these things are in museums and they are there for reasons such as those I have given. But the reasons are seldom indicated. And other very elementary things are commonly missing in the technical accounts of them offered to the visitor. Let me use one as a general illustration. We are hardly ever told what a locomotive cost, to build and to run. True, that may be difficult or perhaps impossible, to find out. In Britain most of the larger railway companies built most, or many, of their locomotives for themselves in their own works, at Crewe or Swindon or Doncaster. The companies' own records do not always allow us to see what the capital cost of those machines was, and if they do it is impossible to attach a really satisfactory figure to an individual machine: it may have been part of a concealed development cost, if the engine is a prototype or comes early in the evolution of a design; it certainly owes a good deal to the very fact that it was built in premises maintained and equipped by the railway company and by staff in its permanent service. If the machine was built by a private manufacturing firm, we shall be lucky to find its records extant; luckier still if they indicate the cost price attributable to the machine and the profit that accrued from making it.

Still, all that said, it is extraordinary how seldom we are told in books or on museum labels how much a locomotive cost. A few people have shown some interest in this matter, notably Professor Saul and Mr Brian Reed, but very few.[8] This is, one might have thought, a basic piece of information about any machine; for machines, after all, are the devices of an economy. The interpretation of the figure, if one has it, may be difficult, but let us at least know what it is.

Similarly are running costs rarely provided: not easy to extract, often impossible, from surviving records, but surely of cardinal importance. If the machine was more economical than its predecessors, that is really worth knowing: a link in a chain of development, perhaps an explanation of what may otherwise be unintelligible such as the continued use of the machine when it was obsolete or inadequate, on the face of it, to its work. So, to take an exceptional instance but a very clear one, the Midland Railway's beautiful express engines with single driving wheels, being notably economical in fuel, were suddenly reinstated in first-class service during the Coal Strike of 1912.

My contention is a simple one, and you can extend it from locomotives to stationary engines or steam hammers or electric cookers if you like. We always need to know as much as we can of the economics of using it. If you think I am labouring something that is obvious, put what I have said to the test. You will probably be surprised by the paucity of the information you are given, until you start to try digging it out for yourself.

A great deal of valuable work has been undertaken on the history of technology, especially since the end of the Second War. We are beginning to see some parts of it much more clearly, to get perspectives and to detect relationships that were hitherto unobserved. The contributions of industrial archaeology and of the new and rapidly multiplied museums I have referred to have been invaluable. To take stock of all this new knowledge and to get the best out of it, we need to do something more than accumulate further knowledge; we need to be confident that we are ceaselessly asking questions arising from it, and that we are getting the right questions. I have suggested one or two, concerned with time and with money. Just as evidently — I am inclined to think more so — the historians must ask themselves questions too, for the government of men and the societies in which they have been grouped are influenced, even regulated, by the technical resources that are available.

We have often been told — though I do not know who coined the phrase — that politics is the art of the possible. That may be truer than we realise. Human beings need management, and the politician has to pay strict attention to what he can and cannot do in that way. He also has to accept limitations on what he can do in a physical or economic sense, which is not perhaps so often consciously understood; it is taken for granted, accepted passively, and that no doubt is common sense. It is for the student of history afterwards to detect the limitations — not only the human ones but also those that are impersonal, many of them technological — and to reckon them up as part of his explanation of what happened, or did not happen, and why. For explanation, after all, in its broadest sense is one of the principal tasks of history.

Notes

1. *Lord Durham's Report on the Affairs of British North America*, ed. Sir C.P. Lucas (1912), ii. 318-9.
2. E.T. MacDermot, *History of the Great Western Railway* (1927-31), i. 620-2.
3. J. Kieve, *The Electric Telegraph* (1973).
4. See G.R. Taylor and I.D. Neu, *The American Railroad Network, 1861-90* (1956).
5. L. Woodward, *The Age of Reform* (1939 ed.), 294.
6. The figures are given conveniently in B.R. Mitchell and I. Deane, *Abstract of British Historical Statistics* (1962), 297-8.
7. See W.B. Baikie, *Narrative of an Exploring Voyage up the Rivers Kwora and Binue* (1856).
8. B. Reed and others, *Loco Profile* (or *Locomotives in Profile*), 36 parts in 3 vols., May 1970-September 1973.

History of Technology in the Teaching of History

R. A. BUCHANAN

There is a potential misunderstanding implicit in the title 'History of technology in the teaching of history' which I would like to clear up at the outset. It could be inferred that I am postulating an essential link between two parts of an equation, as though the teaching of history were incomplete without the history of technology. This is not the case. The relationship between the history of technology and the teaching of history, though useful, is not a necessary one, and to declare or even imply that it is so would overstate the argument. Clio the Muse is a willing servant, always ready to meet the requests of historians for particular types of historical interpretation so that the study of history has come to assume many different forms. Of these, the history of technology is a comparative newcomer seeking to establish itself amongst the range of historical disciplines, and although there are important perspectives which it can bring to the study and teaching of history, it would be wildly presumptuous to claim that it is indispensable. Many traditional historians would dismiss any such claim out of hand, and they are not likely to be impressed by the modest plea that I wish to present. Nevertheless, to carry any conviction, it is important that the case should be presented and open for consideration by those teachers of history who are least likely to be sympathetic to the history of technology.

There is another point of possible misunderstanding which might arise from my title, from which it could be assumed that the history of technology is already a well-established ingredient in history teaching. But in reality the subject has made remarkably little progress towards recognition as a university discipline, especially at undergraduate level. This lack of recognition is particularly the case in Great Britain, where the courses offered by departments of history in the history of technology — except for service courses offered to other departments — can be counted on the fingers of one hand. However, I suspect that the situation is similar in universities throughout the Western World, although the United States probably has a better record than most countries. The fact is that the history of technology is still largely confined to postgraduate studies and research so that any discussion of its role in the teaching of history must refer more to its potential than to its past performance.

I should add that I do not undervalue the service function of the history of technology, especially in providing a historical dimension to the studies of engineers and scientists. On the contrary, as one whose livelihood depended for several years on service teaching, I firmly believe in such studies and I wish that engineers and scientists in positions of educational authority could be convinced of their value. The conclusion may be that an expansion of such teaching is the most promising direction the history of technology can take. But in the context of this paper I see service teaching as essentially peripheral to the subject of relating the history of technology to the main body of historical study and teaching, and at this level we are operating on a very narrow basis of practical experience. I believe several of us have attempted to incorporate units or even substantial sections of technological history in undergraduate teaching programmes. Concerning my own attempts, I am not able to report significant progress and in the present climate of retrenchment in higher education I do not hold out great hopes for the immediate future.

These constrictions on the role of the history of technology are serious, but they present a challenge. Since the subject is emerging as a distinct and articulate sub-discipline of history, it does pose questions of relationships and objectives about modes of study and teaching methods. My plan is to consider the subject of my title under three broad headings. Firstly, I will look at the subject matter of the history of technology in order to distinguish the sort of ideas and concepts that it can contribute to historical understanding. Secondly, I will consider the more practical aspects of the history of technology as a part of what I propose to call 'physical history'. Thirdly, I will seek to answer the question 'What is the use of the history of technology?'. Each of these sections will suggest applications to the teaching of history, and at the end of the paper I will attempt to draw together these various suggestions and try to assess their significance.

History of technology and its contribution towards historical understanding

I have drawn attention elsewhere[1] to the dearth of speculative literature about the history of technology, but the subject is so important and relevant to our present exercise of determining the content of the study that it will bear a brief restatement here. All historical studies contain a largely descriptive element. Indeed, it may be argued that all worthwhile academic studies need a basis of careful and factual description in order to prevent them from becoming too speculative or as jargon-ridden as some social sciences which occasionally seem to possess only the most tentative connection with reality. Because its primary function is the reconstruction of the past, history can never escape from the painstaking research necessary to discover what actually happened at a particular time and place. Such reconstruction may reasonably be described as 'descriptive', although it is as well to remember that selection, interpretation and prejudice enter into even the most simple historical descriptions. The history of technology shares this element in common with other forms of history. As Lord Ashby has reminded us that 'technology is of the earth, earthy',[2] it is to be expected that the quality of factual description will be more pronounced in the history of technology than in those branches of historical research concerned with largely ideological subject matter such as the history of political theory. Most of the very substantial research which has already been published in the history of technology has been concerned with the fundamental task of recording and elucidating particular events or themes in the development of technology, thus establishing a solid core of carefully substantiated factual evidence. In this category may be placed *Oxford history of technology*, the great majority of the papers in the *Transactions* of the Newcomen Society, and most of the standard works on the history of technology that have appeared in the last two decades.[3] It would seem likely that the time has arrived for the history of technology to move on from this firm base and proceed to a stage of interpretative and speculative evaluation which will relate the parts to each other and to general principles of organization. It would also seem likely that the subject can now begin to feed some ideas into the teaching of history.

As a preliminary framework of analysis, it is possible to fit most of the descriptive material in the history of technology into one or other of three main categories: innovation, where historians are well equipped to discuss the social, technological and psychological prerequisites of successful invention; technological transmission, where the nature of the process is a legitimate subject of historical examinations; and the socio-political impact of technological developments, which has already been considerably debated. These three areas of analysis correspond to the beginning, the spread and the consequences of innovative processes, and they indicate the centrality of innovation to the history of technology. It would not be appropriate to give a full account of the scholarly activity in these fields here, since it would involve a survey of all the most interesting work in the history of technology in recent decades. It will be enough to give a brief summary of the scope of each area in order to determine the contribution each may make to the teaching of history.

In the first area, pioneering work was done in the analysis of the sources of invention in the inter-war years by Usher and Schumpeter.[4] Their definitions and distinctions between such terms as invention, innovation and development have subsequently been refined and modified in such studies as those by Jewkes *et al.*[5] and by Langris *et al.*[6] while Schmookler has shown how the economic analysis of patent registrations can be used to extend our understanding of the inventive process.[7] Comparatively little has been done to probe the psychology of inventors, although the phenomenon of 'simultaneous invention' has attracted attention and led some historians of a determinist ideology to maintain that the economic and social circumstances are more important in the generation of invention than the genius or serendipity of individuals. Samuel Lilley came close to this view in his popular work, *Man, machines and history*, first published in 1948, although he subsequently dropped his discussion of the measurement of technological progress, with its suggestive 'relative invention rate' index, from the revised edition in 1965.[8] An inversion of the analysis of the cause of invention has led historians to consider the lack of invention in the classic civilizations of Greece and Rome and to speculate on the role of social factors such as the existence of slavery as a cause of technological lethargy and thus also of the decline of Hellenic civilization.[9] Conversely, again, the invention of the mechanical clock and of printing by movable type has stimulated discussion of causes and consequences which are partly technical but are more concerned with the personal and social factors involved.[10] Despite all this wide-ranging interest in the sources of invention, there remains a wide margin of disagreement regarding the causal priorities and the subject of invention requires continuing research of a highly analytical content.

Technological transmission, the second area of analysis, has also provided ground for discussion. One of the great academic debates amongst archaeologists and ancient historians has been concerned with the divergence of interpretations about the spread of technological innovations in prehistory and early civilizations between the 'diffusionists' who have related similar inventions to a single source and the supporters of 'spontaneous innovation' who have preferred to emphasize the independence and regional peculiarities of such innovations. The debate remains open-ended, even amongst archaeologists who have argued it at great length, and intrepid experimentalists have been prepared to cross the Pacific and Atlantic Oceans on primitive rafts in order to test the feasibility of the diffusionist thesis.[11] More recent historical periods have generated their own scholarly problems of technological transmission. For instance, the substance and manner of the transmission of the windmill and of gunpowder from the Far East to Western Europe in the Middle Ages have been fertile subjects for speculation,[12] while in modern times the spread of steam technology from Britain to continental Europe,[13] and of European technology to America (and vice versa)[14] have received scholarly attention. Maurice Daumas, as the first Secretary-General of the International Committee for the History of Technology (ICOHTEC), wisely directed the deliberations of this putative effort at international co-operation in the history of technology towards the relationship of the initiator to the recipient in the transmission of technological innovation from one country to another.[15] Clearly this is an important subject for academic analysis.

The third area of analytical scholarship in the history of technology is rather more diffuse than the other two and more difficult to present in a systematic form. But there can be no doubt that technological innovation is a potent agent of social change so that it is useful to be able to indicate how it operates. Several lines of investigation, all potentially fruitful, may be observed. Technological innovation disturbs 'natural' population mechanisms, for example, by enabling more food to be produced from the same acreage and by removing causes of disease.[16] It disturbs 'normal' social arrangements by creating a need for a large, controlled labour force with factory discipline,[17] and by advances in military technology which alter political balances of power and encourage war.[18] It makes possible and positively promotes urbanization.[19] It disrupts 'barriers' of space and time by improving transport and communications.[20] It disturbs existing value systems and modifies artistic and literary images.[21] Historians of technology have begun to apply themselves to all these cultural consequences of technological innovation, albeit somewhat sporadically and not always effectively. The columns of *Technology and Culture*, in particular, have performed a valuable service in providing a forum for this widening discussion about the history of technology. One recent issue, for example, contained a collection of illuminating articles by Robert P. Multhauf, Edwin Layton, Eugene Ferguson and Derek de Solla Price.[22] Another had a provocative feature by Reinhard Rürup on 'Historians and modern technology — reflections on the development and current problems of the history of technology'.[23] All these are promising indications of work in progress, but much remains to be done before anything resembling an agreed interpretation of this complex set of problems can begin to emerge.

The definition of these areas of examination may appear simplistic to scholars who, in approaching the history of technology, use highly conceptual models of a mathematical or sociological nature which have recently been helpful in elucidating some aspects of economic history.[24] But the matter-of-fact quality of technology — its 'earthiness' — is a vital characteristic of the history of technology so that it is important to avoid over-sophistication in the analysis of its features. Even though its ideological relationships and implications are important, they are rarely paramount and it is a mistake to impose a dogmatic conceptual structure or model on the history of technology. There must of course be conceptualization and model-building in the subject, which has certainly suffered in the past from a lack of analytical thought, but the danger of overstating the case for remote mathematically based abstractions or bewildering sociological terminology should be resisted. Where there is a need for mathematics, it is for arithmetic rather than algebra or the higher calculus. The history of technology has admittedly suffered from a lack of systematic quantification in the past. One consequence has been what might be called 'the fallacy of immediate invention' whereby it has been assumed, for lack of an adequate statistical basis of generalization, that a good idea is more or less automatically adopted once it becomes available. Hence the value of studies such as those which have endeavoured to calculate in detail the numbers of Newcomen-type steam engines which remained in operation after the introduction of James Watt's improvements and which have demonstrated the persistence of the older technology long after better machines came on the market.[25] It is astonishing, on reflection, how little statistical or tabular information is presented in the five volumes of the Oxford *History of technology*; the omission indicates the magnitude of the research tasks still awaiting attention in this subject. But a recognition of the need for adequate quantification is not an admission that the history of technology needs to achieve an abstruse level of abstraction. On the contrary, a firm statistical base is necessary in order to derive more accurate inferences from technological innovations, transfers and impact that any made so far.

From the discussion so far it should be clear that I am advocating a scholarly *via media*. By applying analytical and statistical techniques to certain critical areas, my approach lifts the history of technology above the descriptive account where one thing happens after another. On the other hand, my approach leaves the subject firmly on the ground so that the discipline is not made subservient to the regimen of empty conceptual boxes which have to be filled with the available data. It seems to me that this *via media*, which is central to the purpose of my paper, is most likely to be relevant to the teaching of history. For while the scholar and teacher of history can safely ignore the descriptive, factual monograph on the technical performance of engines, machines and processes which has been the mainstream of publications on the history of technology until recently, and while he can equally safely ignore the refined mathematical abstractions of those who would over-conceptualize the subject given the opportunity, the systematic treatment of the key themes which I have indicated can be of immediate value to a wide variety of historians. Economic historians especially have not been slow to appreciate the value of the history of technology in their own assessments of such problems as the lag in the transfer of innovations, as shown by the work of Habakkuk, Saul, and Rosenberg.[26] Social historians are equally receptive to attempts to interpret the cultural impact of technology and political historians recognize the significance of technology in warfare. These are sure signs of the ability of the history of technology by the patient elucidation of its own central themes to make a significant contribution to the teaching of history in many sub-disciplines beyond its own frontiers. The recognition of this contribution depends partly on the practitioners of the various sub-disciplines concerned but more on the persistence of historians of technology in offering interpretations, concepts and ideas which illuminate a wide field of historical study.

History of technology as part of physical history

In addition to its conceptual contribution, both actual and potential, to historical scholarship, the history of technology can also make a highly practical contribution which is particularly relevant to the teaching of history. In this respect, the history of technology is a species belonging to the genus of historical disciplines which may be called 'physical history'. Together with local history, architectural history, archaeology, historical geography and several other types of study, the history of technology cannot avoid an emphasis on field-work. R.H. Tawney reputedly first called on historians to invest in a stout pair of walking shoes and W.G. Hoskins took up the call in his work on the English landscape.[27] Michael Rix in launching the study of industrial archaeology went further, requiring of practitioners a good pair of gumboots.[28] What all of these spokesmen for physical history agree is that the physical evidence available through field-work to the senses of sight and touch has a vital contribution for historical scholarship. It may not be as important as documentary evidence, especially in areas of modern history where there is a surfeit of documentation and in such subjects as diplomatic history or the history of political thought where physical evidence is elusive and peripheral. But where written evidence is lacking or inadequate—and these conditions pertain for our knowledge of all periods except the most recent — physical evidence, properly recorded and judiciously interpreted, can provide supplementation, while in such areas as the archaeology of prehistoric societies it becomes paramount. Moreover, in the most recent periods of history when there is already too much written material to assimilate, physical evidence can supply a missing dimension. Most particularly, it can give the student a sense of identity with the past that no amount of book-learning can inculcate. At the very least physical history is a useful teaching aid to the historian. And amongst the evidence examined by physical history and thus made available as a teaching aid to all historians are the surviving artefacts of past technologies, which it is the business of the historian of technology to interpret.

It could be argued that the preoccupation with the surviving evidence of obsolete industries and transport systems has acquired the status of a sub-discipline in its own right under the name of 'industrial archaeology'. As you might expect, I would go some way towards arguing this myself. But it is a matter of drawing a highly arbitrary line to distinguish so sharply between industrial archaeology and the history of technology. Admittedly, in so far as historians of technology are engaged in the important and overdue exercise of increasing the analytical and conceptual quality of their discipline in the ways already advocated, they have tended to show somewhat less propensity for field-work in their subject, and to this extent there has been in the last decade a certain drawing apart of industrial archaeology from the history of technology. However, a complete separation would be harmful to both and I prefer to regard them as complementary aspects of the same exercise, both being related in turn to the other subsections of physical history. I wish to make the point, therefore, that through industrial archaeology and physical history, the history of technology has a very real contribution to offer to the study and teaching of history.

Information useful to the historian can be derived from physical evidence at four levels. Firstly through the study of landscapes regarding such areas as geology, drainage, soil conditions and mineral resources and using a variety of specialized techniques such as aerial photography, it is possible to learn how any one particular landscape has been modified by man and so add to our understanding of its history. The identification of forgotten prehistoric sites through crop marks or of deserted medieval villages through hedge patterns are both examples of highly rewarding research of this nature in recent years.[29] The second level is concerned with existing settlement patterns, including farms, villages, small towns and cities, and the corresponding transport systems. Urbanization may be regarded as the dominant settlement pattern of a fully industrialized society and the study of the physical features of towns can elucidate the process of urban growth. The history of the Victorian town, for example, has benefited substantially from such examination.[30] Thirdly, structures such as houses, factory buildings, mills and public utilities provide further level of physical enquiry. Here the architectural historian has most to contribute, but as industrial buildings are usually highly functional, the historian of technology is often best qualified to interpret their shape and layout.[31] Finally, the artefacts in the shape of machines and equipment, street furniture and commercial products are the fourth level of examination for the physical evidence, where the history of technology has most application. In interpreting the remains of an obsolete metallurgical process through its surviving artefacts, the skills of the materials scientist are invaluable, and for interpreting machinery of any age some engineering competence is equally useful. These are skills which the historian of technology usually knows where to find even if he does not possess them himself.

Perhaps I may permit myself a single illustration of this point in an area where I have found particularly exciting research possibilities. The lifetime spent photographing and recording stationary steam engines by George Watkins has succeeded in transferring information about a rapidly disappearing species of artefact to a permanent archive. As such, it is available to scholars and serves as a source of inspiration and enthusiasm while also raising a series of provocative questions for further examination. How many steam engines were produced of different types and for different services in the nineteenth century? What factors influenced the continuing improvement in the performance of steam engines throughout the nineteenth century? Why was there an apparent 'lag' in the application of improvements in some services such as marine steam engines rather than in others? What influence did the development of steam technology have on the internal combustion engine and how did steam technology react to competition from this and other alternative sources of power? For these and many other questions no precise answers are as yet available. However, the fact that they are being formulated as topics for research indicates the powerful stimulus derived from a painstaking attempt to collect and to organize systematically the physical evidence of steam technology by a scholar with the engineering competence to understand what he was observing while the artefacts were still available for study. This undertaking by Watkins also provides a valuable teaching aid for engineering and history students alike in learning to appreciate the qualities of steam engines and the consequences of the power revolution which they brought in the growth of an industrial economy.[32]

This outline of the sort of information which can be acquired from physical history demonstrates the indispensable contribution that the history of technology can make, especially at the levels of structure and artefact analysis. This contribution is primarily of a scholarly nature: like any historian, the historian of technology is concerned with the re-creation, as exactly as he can make it, of past events, using as his distinctive technique in this exercise the interpretation of artefacts, whether these be recorded in documentary form or available for examination in three-dimensional reality. The secondary contribution is pedagogic, because the account of how a machine works or an artefact functions can explain more convincingly and teach more effectively than a corresponding paper exercise. The practical aspect which the history of technology can give to the processes of instruction in this way has so far been little understood, despite efforts by some of our more progressive museums to apply the lesson. It is small wonder therefore that historians have been slow to take advantage of these pedagogic opportunities although there are encouraging signs of attempts to combine instruction with field-work and visits to sites of industrial archaeological significance. There are great educational possibilities for this development which such schemes as 'Project Technology' are just beginning to exploit.[33]

What is the use of the history of technology?

Our treatment of the conceptual and practical aspects of the history of technology does not exhaust its possible contribution to the teaching of history. Another aspect, with teaching implications, is suggested by the question, 'What is the use of the history of technology?' The German scholar Dr. Rürup, in the article in *Technology and Culture* already cited, argued the study of the history of technology has a part to play in the vigorous contemporary debate about the role of technology in society and the problems involved in the relationships between them, 'for historians are in a position to contribute significantly to solving such problems by applying their discipline to studies of the changing relationship between technology and society, the specific concrete causes and effects of technological progress, and the changing relationships between technology and science'.[34] This essay is valuable for its insight into the function of technological history and for its survey of the existing state of the discipline, which is refreshingly internationalist in its scope. But Dr. Rürup does not give any very precise indications of the ways in which the history of technology can illuminate the 'technology and society' debate. I would like to suggest that there are some specific issues in this debate to which the history of technology can make a contribution and that in doing so, the subject acquires a practical utility which has consequences for the teaching of history.

In supporting Dr. Rürup's contention that the history of technology can make a useful and, indeed, an essential contribution to the contemporary debate about the role of technology in society and the closely related issues of economic growth, ecological balance, pollution and conservation, I am aware that I am going beyond what many professional historians believe to be the legitimate boundaries of their subject. No one would deny that there are strict limits to the applicability of historical scholarship to this debate. Every discipline has such limits, but those of history are particularly obvious because it deals exclusively with the past, whereas the main thrust of the debate is in current affairs and future policy. However, in so far as our understanding of past experience is relevant to our comprehension of the present and future, the historian may make comparisons and produce insights which can be of immense value in making judgements about the present and projections about the future. The word used, it is necessary to emphasize, is 'projections' rather than 'predictions': the predictive quality after which the other social sciences sometimes hanker is not the business of history, but by inference and analogy from experience of past situations the historian is able to suggest patterns of relationships which are likely to recur, and such patterns are implied by the word 'projection'. In this way the historian can draw on the past for guide-lines to the present and future and the historian of technology in particular can suggest relevant projections on several critical issues in the contemporary debate. To the teacher of history who is anxious to justify his discipline in terms of its utility and relevance the history of technology is an especially rewarding field of enquiry. To all other historians what I have to say next will have no significance.

The contribution which the history of technology can make to the contemporary debate about technology and society falls precisely into those areas in which we have already observed the subject to be acquiring analytical depth. Firstly, there is much of value to be interpreted from historical experience about the process of innovation. Studies of innovation in the Industrial Revolution in Great Britain, such as the essays collected by Professor A.E. Musson[35] have a relevance to the general question of whether invention can be artificially stimulated and, if so, how. It is possible to argue, for example, that one of the essential prerequisites for the inventive boom in eighteenth-century Britain was the presence of a socially significant middle-class group which was disposed to experiment with new ideas and to apply them when they had any promise of personal profit. Such a historical interpretation may then be brought to bear on the examination of a currently developing society and may help to determine policies aimed at encouraging similar social focal points of enterprise in these communities. It is barely necessary to stress the tentative nature of such a projection, when so many other factors need to be taken into account simultaneously, and in any event it implies judgements about the desirability of particular patterns of development which the historian may consider to be beyond his brief. But at the very least, the historical experience provides a clue about one outstandingly fertile set of relationships which, along with other clues, may contribute to the solution of one aspect of a complex contemporary problem.

Secondly, the analytical area concerned with assessing the nature and scope of the ways in which technological innovation has been transmitted in the past from one society to another can supply helpful projections for the modern politician and administrator. The role of skilled personnel and professional journals, for example, in transmitting European technological experience to America early in the nineteenth century can suggest ways in which the dissemination can be continued to parts of Asia and Africa in our own time. Again, the differences between these situations are obvious enough, but the diligence of the historical scholar in explaining the way in which the new textile technology crossed the Atlantic at the turn of the eighteenth and nineteenth centuries can indicate possible processes of diffusion for modern electronic technology: this example can be reinforced by countless other pieces of historical experience. In *The maze of ingenuity*, Arnold Pacey recently attempted to draw out the application of lessons learnt from the process of Western industrialization to the developing societies of the world today. Although not completely satisfactory in the fulfilment of its objective, this was a stimulating and pioneering study and Pacey subsequently abandoned his academic career to promote 'intermediate' and 'alternative' technology through Oxfam. His action is a very pertinent indication of the practical implications of the history of technology.[36]

Thirdly, in the analysis of the relationship between technology and culture, the historian of technology can show insight into the transformation of our own society and suggest policies for assisting or resisting similar changes in other societies. We have already observed that the implications of this interaction between the technological and cultural aspects of social life are extremely complex so that virtually no policy can be guaranteed to produce the desired effect and no other. The current debates amongst town-planners about policies to relieve traffic congestion and about 'high rise' development indicate the largely contingent and pragmatic quality of these policies, isolated from historical roots, so that any historical experience may help to determine the balance between conflicting possibilities. The work of historians of technology in examining the development of urban services can thus have contemporary relevance, as can work on the history of automatic controls and technological education. Even more important, at least potentially, is the work of analysing the impact of technology on value systems and its contribution to such key-concepts as 'progress'. The technological aspects of the 'ecology crisis' also have an important historical dimension, and as shown recently by Richard Wilkinson,[37] the role of technology in this area may be more ambiguous than has been commonly supposed. Amongst other provocative ideas, Wilkinson indicates a fallacious tendency to equate quality with quantity in the implication that because an innovation produces more, it is therefore better. It is frequently only different, being a cheaper substitute for whatever it replaces. For example, when coke-smelting of iron replaced charcoal-smelting in eighteenth-century Britain, the former not only used a cheaper and more readily available fuel but it also promoted the construction of larger blast furnaces so that more iron could be cast. But the quality of coke-iron remained inferior to charcoal-iron even though the change became widespread. The implications of this and other historical resource substitutions for our understanding of ecological disequilibrium and economic growth are significant and have an important bearing on many topical policy decisions in relation to the developing countries.

Enough has been said, I hope, to suggest that there are certain very specific issues on which the history of technology can make something more than a casual contribution to the contemporary discussion ranging over all aspects of the relationship between technology and society. This contribution is an important function of technological history and arises from the main analytical preoccupations of the subject. As such, it has relevance to the teaching of history in two ways. First, it gives a specifically utilitarian orientation to the teaching of the history of technology because it emphasizes the value of a historical understanding in tackling a wide range of contemporary problems. The subject therefore becomes an important adjunct to teaching on environmental studies, ecology, the energy crisis and so on. Second, it provides all professional historians with an example of the practical possibilities of their discipline and could inspire them to make linkages and interpretations relating their historical experience to the problem of understanding the complex patterns of motivation which provide the dynamic force of our civilization. Such an example will only be followed, needless to say, by historians who are already conscious of a need to demonstrate the utility of their discipline. But those are the historians who are most likely to take seriously the history of technology.

Summary

I have been arguing that the history of technology has an important place in the teaching of history. By sharpening its own conceptual and analytical focus it can contribute some important interpretative ideas to history in general and such ideas are already influencing parts of economic and social history. By exploiting its intimate relationship with industrial archaeology as an aspect of physical history, it can demonstrate the value of technological structures and artefacts in illuminating many parts of history for which documentary records are inadequate, while at the same time providing superb teaching aids to help the comprehension of students. By applying its insights to problems of the contemporary world, it is possible to justify the use of historical studies in general as well as of the history of technology in particular and to achieve a sort of functional relevance to the teaching of history which is at present rare. In the last resort, it seems to me that as we strive, as strive we must, to perfect our craft as historians of technology, we should never forget its essential earthiness and humanity. The history of technology is about people, and the way in which people have made and done things, and the implications of these actions upon each other. It is this human content to the subject which gives it valuable relevance to the great contemporary questions of our civilization and makes it important in the teaching of history.

Notes

1. Angus Buchanan, 'Technology and history', *Social Studies of Science*, 5 (1975) 489-99. Essay review.
2. Eric Ashby, *Technology and the academics*, London, 1958, p. 66.
3. C. Singer *et al*, (eds.), *A history of technology*, 5 vols., Oxford, 1954-8. The Newcomen Society for the Study of the History of Engineering and Technology has published its *Transactions* annually since 1922. For a succinct review of the standard works on the history of technology, see Robert Multhauf, 'The historiography of technology: some observations on the state of the history of technology', *Technology and Culture*, 15 (January 1974), 1-12.
4. A.P. Usher, *A history of mechanical inventions*, rev. ed., Cambridge, Mass., 1954; J. Schumpeter, 'The instability of capitalism', *Economic Journal*, (1928), 361-86; and useful extracts are given in Nathan Rosenberg (ed.), *The economics of technological change: selected readings*, Harmondsworth, 1971.
5. John Jewkes *et al.*, *The sources of invention*, London, 1958.
6. J. Langrish *et al.*, *Wealth from knowledge: studies of innovation in industry*, London, 1972.
7. Jacob Schmookler, *Patents, invention and economic change: data and selected essays*, Cambridge, Mass., 1972.
8. Samuel Lilley, *Man, machines and history*, 1st ed., London, 1948; rev. ed., London, 1965. The book loses conviction because of its blatant Marxist and pro-Soviet bias.
9. See for example M.I. Finley, 'Technical innovation and economic progress in the ancient world', *Economic History Review*, 2nd ser. 18 (August 1965), 29-45.
10. The relationships were lucidly presented in Lewis Mumford, *Technics and civilization*, London, 1934.
11. Thor Heyerdahl, *The Kon-Tiki expedition*, London, 1950 and *The Ra expeditions*, London, 1971. Both available in paperback.
12. See for example Lynn White Jr., *Medieval technology and social change*, Oxford, 1962, especially chapter 3.
13. Amongst recent contributions to this discussion, see G.J. Hollister-Short's paper to the Newcomen Society, 'The intoduction of the Newcomen engine into Europe' (November 1976). A more general treatment may be found in S.B. Saul, 'The nature and diffusion of technology' in A.J. Youngson (ed.), *Economic development in the long run*, London, 1972.

14. The classic treatment of this subject is H.J. Habakkuk, *American and British technology in the nineteenth century*, Cambridge, 1962.

15. Published in '*L'acquisition des techniques par les pays non-initiateurs*' in *Colloques internationaux du Centre National de la Recherche Scientifique*, Paris, 1973.

16. This effect is well-presented in R.G. Wilkinson, *Poverty and progress: an ecological model of economic development*, London, 1973.

17. See for example the analysis in E.P. Thompson, *The making of the English working class*, London, 1963, particularly chapters 6, and 11.

18. See for example contributions to the *New Cambridge modern history*, e.g. Michael Lewis, 'Armed forces and the art of war: navies' (Chapter XI) and B.H. Liddell Hart, 'Armed forces and the art of war: armies' (Chapter XII) in Volume X, *The zenith of European power 1830-1870*, Cambridge, 1971.

19. A theme developed in Lewis Mumford, *The culture of cities*, London, 1938.

20. See for example the perceptive appreciation of the impact of the railways on society by a leading social historian in Harold Perkin, *The age of the railway*, London, 1970.

21. This theme was investigated by F.D. Klingender, *Art and the Industrial Revolution*, 1st ed., London, 1947; rev. ed., by Sir Arthur Elton, London, 1968. See also Herbert L. Sussman, *Victorians and the machine, the literary response to technology*, Cambridge, Mass., 1968.

22. *Technology and Culture*, 15 (January 1974), 1-48. The papers were delivered to a symposium of the Society for the History of Technology (SHOT) in Washington, D.C. in 1972 under the general title of 'The historiography of technology'.

23. Reinhard Rürup, 'Historians and modern technology — reflections on the development and current problems of the history of technology', *Technology and Culture*, 15 (April 1974), 161-93.

24. I refer to the use of such methods in the so-called 'new economic history' by American authors in the first instance, but with some following amongst British economic historians. For a representative selection see Peter Temin (ed.), *New economic history — selected readings*, Harmondsworth, 1973.

25. See A.E. Musson and E. Robinson, *Science and technology in the Industrial Revolution*, Manchester, 1969, especially Chapter 12, 'The early growth of steam power', a version of which originally appeared in the *Economic History Review*, 2nd ser. 11 (1959).

26. See Habakkuk, *American and British technology* and Saul, 'Nature and diffusion of technology'. See also S.B. Saul (ed.), *Technological change: the United States and Britain in the nineteenth century*, London, 1970 and Rosenberg, *Economics of technological change*.

27. W.G. Hoskins, *The making of the English landscape*, London, 1955 called in his introduction for 'a combination of documentary research and of field-work, of laborious scrambling on foot wherever the trail may lead' (p. 14), and in his *Local history in England*, London, 1959, he claimed that no historian 'ought to be afraid to get his feet wet' (p. 2).

28. M. Rix, 'Industrial archaeological field work' in R.A. Buchanan (ed.), *The theory and practice of industrial archaeology*, Bath, 1968, p. 19.

29. Hoskins, *Making of the English landscape* pioneered work in this field. Amongst more specialized studies are M.W. Beresford and J.G. Hurst (eds.), *Deserted medieval villages*, London, 1971 and M.W. Beresford and J.K. St. Joseph, *Medieval England — an aerial survey*, Cambridge, 1958.

30. There is remarkably little physical history in H.J. Dyos and M. Wolff (eds.), *The Victorian city — images and realities*, 2 vols., London, 1973. But other recent urban studies are more aware of the physical reality, e.g. J. Hume. *The industrial archaeology of Glasgow*, Glasgow, 1974. And even such a general work as Asa Briggs, *Victorian cities*, London, 1963 conveys a strong sense of time and place by its perception of physical differences.

31. Jennifer Tann, *The development of the factory*, London, 1970 is a good example of this approach. Also K.C. Barraclough, 'The development of the cementation process for the manufacture of steel', *Post-Medieval Archaeology*, 10 (1976), 65-88 is an excellent illustration of skill in the interpretation of physical evidence.

32. This theme has been explored in R.A. Buchanan and George Watkins, *The industrial archaeology of the stationary steam engine*, London, 1976, and a detailed application to the questions raised is now being worked out in the subsequent research programme at the Centre for the Study of the History of Technology, University of Bath.

33. 'Project Technology' has produced some enterprising schemes for using technological history in secondary and technical education.

34. Rürup, 'Historians and modern technology', p. 166.

35. A.E. Musson (ed.), *Science, technology and economic growth in the eighteenth century*, London, 1972.

36. Arnold Pacey, *The maze of ingenuity: ideas and idealism in the development of technology*, London, 1974. See also the essay review of the book in Buchanan, 'Technology and history'.

37. Wilkinson, *Poverty and progress*.

The Role of History in a Civil Engineering Course

P. B. MORICE

It may be timely, I believe, to look again at the place of the history of engineering and technology in the civil engineering course. This paper is thus in no way a contribution to the history of engineering for it is concerned with historiology rather than the history itself.

In the past ten to fifteen years, with the growth of major engineering works of unprecedented scale, there has developed a widespread concern for the conservation of the environment and its amenities. The post-war popularity of archaeology now seems to have shifted discernibly towards industrial archaeology and those artefacts which have been largely produced by engineers. At the same time, a major debate is developing over the education of engineers and technologists. This debate must surely lead to a revision of the basic engineering course, which will cause the young engineer to be more aware of his public accountability in environmental matters.

In so far as civil engineering is a profession, it represents a received body of hard-won knowledge which takes a considerable effort to acquire and which is almost totally dependent upon the work and contributions of those who have gone before. Although much of engineering is based upon science and therefore theoretically capable of development from first principles, it is not always possible or even desirable to establish totally new solutions and it is often the unquantifiable imponderables which account for the success or failure of a civil engineering enterprise. The body of civil engineering knowledge is inherently a historical package or heritage, the historical content of which it is perilous to ignore. The history of civil engineering is littered with examples of failures which historical insight could probably have avoided. But then one could quote Hegel on the affairs of the human race that 'the one thing one learns from history is that nobody ever learns anything from history'.[1] Hegel is suggesting that while there are things to learn from history, most of us never bother to do so. A more conscious study of engineering history might enable us sometimes to avoid suffering the condemnation attributed to King Alfred that 'I know of nothing worse of a man than that he should not know'.

In some ways one can separate the engineer's concern with history into two parts: the first satisfies the personal interests of the individual, while the other enables him to perform better as a professional engineer.

Personal interest is of course a strong motive for doing anything and it is perhaps a natural curiosity in all of us to want to know how others have coped with their problems, particularly those similar to our own. It would seem strange for an engineer to be totally disinterested in this aspect of his own subject and the civil engineer perhaps above all might be expected to think this way since he is surrounded by so much of the evidence of his predecessors' activities. The built and engineered environment may be changing more rapidly than ever before. However, no society since the tented nomads can replace more than a small proportion of its built environment within one man's working life. Thus the history of civil engineering is all about us and provides an almost inexhaustible scope for study and amusement, leading to a deeper understanding and interest in one's own professional activities. Few can surely remain unmoved or disinterested once they become aware of the situation. I am personally delighted that a dry dock of 1698 which I can see in Portsmouth Dockyard is still in use.

In a civil engineering course, therefore, some history of the subject should provide an opening to what Dr. Hamilton has called the 'cultural field of one's own profession'.[2] From many discussions with colleagues and friends who have made significant contributions to engineering historical studies, I believe that this is probably the principal motivation, which can also be satisfied by the passive as well as the active participant. There is much written material for those who care to search it out.

The professional implications are perhaps more widespread. Above I have suggested that the whole of the transferred body of knowledge is in a sense also its history since essentially it is handed down to each new member of the profession, but there is more to it than this. One way, which, although clearly not the only way, is perhaps one of the better, is to study a logical or analytical solution to a problem in the context of its development. Theories rarely come out of thin air and are only exceptionally complete and satisfactory in their first explanations. Newton's laws of motion may have been the result of a man of genius putting his mind to a problem but he himself said that 'if I have been able to see a little further it is because I have stood on the shoulders of giants'.[3] Any study of classical mechanics will show the evolution of ideas probably first chronicled by Aristotle and then refined during the years by Bradwardine, Buridan and Oresme through to Copernicus and Galileo until the threads were finally pulled together into a coherent theory.[4]

My beliefs are that necessity is the mother of invention and that theories evolve. I am not surprised that two or more people may come up with the same solution to a problem at the same time, as for example Newton and Leibniz with calculus; I am, however, surprised that it should be so often supposed that only one man has done so and consequently that so much effort should be devoted to establishing a definitive attribution to the 'inventor'. If one takes pre-stressing as an example, most people will say that Eugène Freyssinet was the inventor. There can be little doubt that he developed and successfully applied prestressing to concrete in civil engineering after a painstaking period of trial and error. Although it is not clear how much Freyssinet was consciously aware of the work which had gone before, it was perhaps the technical environment of the times, unconsciously perceived, which was the more important factor in influencing and stimulating him.

As Sutherland has shown,[5] engineers in the 1840s were concerned to find some way of overcoming the poor tensile qualities of cast iron, then a relatively new structural material, and felt they had found the solution in wrought iron. The casting of wrought-iron reinforcement into cast-iron beams had been tried, but the idea of pre-stressing had also been developed to overcome this material deficiency. Techniques very similar to those used today were adopted involving the idea of joining together pre-cast units to form the whole structure. The value of a varying eccentricity to follow the bending moment seems to have been understood, but not the need to maintain the centre of pressure within the core of the section. Thus, while tensile stresses were eliminated in mid-span, they were actually artificially increased close to the supports and consequently were a major factor in inducing failure.

We can look much further back in time to see another use of pre-stressing when the tensile jointing of timber planks was unsatisfactory. The Egyptian shipbuilders of 2600 B.C. needed to build long ships with short planks, which were the only ones available, and in particular to provide a large enough cantilever at bow and stern for loading at river banks where there were no quays or jetties. The natural hogging moment thereby induced was counteracted by an artificially produced sagging moment, which furthermore could be adjusted through varying the pre-stress force by twisting the rope cable; in the same way the carpenter's bow-saw blade was pre-stressed against buckling. Even such a brief look at these historical aspects of pre-stressing can throw light upon some of the problems of its development and its nature as a technique.

This example is perhaps concerned with a special and unusual situation where the engineer is trying to find a solution to a problem. To say there is nothing new under the sun is too glib for the modern world is clearly possessed of a vast range of entirely new products and artefacts. It is perhaps the conjunction of ideas and technical possibilities which yields the new; often the idea has been around for a long time waiting for the means of application. It appears that Charles Babbage and Lady Lovelace had at least the idea of the automatic digital computer in the last century. They failed to produce a satisfactory machine because the only means available were mechanical rather than electronic.

To move to my second belief that theories evolve, I believe that there are many occasions when the civil engineer may make use of the historical approach to assist in carrying out his tasks, although I have no substantial body of evidence to support this. The repair, maintenance and modification of old structures and works would be greatly facilitated by adequate documentary evidence of previous enterprise; reports of difficulties which had to be overcome and the ways in which the works were conceived and constructed might also be helpful. Where such specific evidence is lacking a knowledge of contemporary practice can usefully substitute. Even in the construction of new works an informed study of the site and previous works can provide much useful information and perhaps concentrate an area of study such as soils investigations.

My last plea for the civil engineer's study of history is in response to his increasingly recognized responsibility for the environment and cultural heritage. Modern man may not be able to replace everything about him in his lifetime, but when he does build or rebuild, his efforts are likely to be significantly more disruptive than in former times and can easily result in the total destruction of all archaeological evidence. Motorway construction is a prime example. The 150-foot wide slice made through the landscape will certainly destroy what may be the record of two or three thousand years of previous occupations. Enlightened organizations have enabled the civil engineer to be joined by the professional archaeologist and together they have organized digs to fit construction programmes. However, much evidence will be revealed by the preliminary earthworks when the archaeologist may be off the site; the civil engineer will need to be aware of the type of finds to be expected and to be able to recognize the unexpected before it is destroyed. Only a genuine interest and concern and a modicum of real knowledge can bring this result. When the M3 was being constructed in southern England, an average of two occupation sites for each mile of motorway was discovered, making a total of eight-four, and in fifty square miles of Wessex something in excess of 400 sites are to be expected.

Another aspect of this principle of responsibility is the care of objects and records. Many civil engineers will find themselves the trustees of the works of their predecessors and it is a responsibility vested in them not to neglect but to respect this trust. When past structures reach the end of their useful lives, the choice lies between preservation and demolition. It is a principle of industrial archaeology that it is neither possible nor desirable to litter the countryside with useless memorials to past ages. However, it is reasonable to preserve some prime, typical or unique examples and encumbent upon the engineer to know enough of his subject to be able to judge the value of each object and make appropriate recommendations. This trust is particularly important for drawings, and details and specifications; documents of use to future researchers. It takes little effort to preserve and perhaps make photographic records of documents and where none exist or are inadequate, to prepare measured drawings of the objects before demolition takes place.

I have outlined a number of reasons for the study of the history of civil engineering by the informed civil engineer. How should such studies be fitted into the undergraduate civil engineering course in a stimulating and not superficial way? I am against the rapid slide show of large numbers of pretty pictures of remarkable structures, because we should be concerned to explain the how and the why. The relevance of history can be made clear by teaching how particular ideas came about and were adopted in theory and practice as in the case study method.

The simple theory of beams as developed by Galileo, Mariotte, Jacob Bernoulli and Thomas Young seems an interesting and instructive way to introduce students to the elastic theory. It enables one to point out how much can still be achieved when a theory is not yet complete, as when Euler could establish an instability theory for columns without a correct stress distribution. At the same time the engineer's responsibilities can be explained by following the development of a particular technology, such as cast iron in buildings or the establishment of railways, and studying its influence upon the subsequent history of society. Such a study could become a large task indeed, but a vignette of the subject can be very illuminating. One can also use the student's development of the technical skills of surveying and drawing to record works and buildings. At a later stage in the course the lead of Heyman[6] and Mainstone[7] could be followed in the analysis, using modern methods, of past structural achievements and, for example, the problems cause by settlement.

Many degree courses encourage the development of individual curiosity through project work, usually in the final year. A few well-motivated students can profit by directing their projects towards historical subjects. A recent undergraduate of mine happily and successfully engaged in the study of the maritime piled pier which seemed to have been a neglected area.[8] He was able to unearth a great deal of information and put together a thesis which closely linked the maritime piled pier with the development of steam navigation, the railways and the Bank Holiday Act of 1871.

I find that many young engineers are stimulated to realize that they are in the direct line of the great names of the past and to this extent the cult of the achievements of the individual engineers of history can be advantageous.

One must end by saying that, clearly, all these objectives are unlikely to be achieved in the present overcrowded three-year course and in the recent past the rapid pace of technological progress in all branches of engineering has tended to oust the more contemplative study of history. However, we are becoming increasingly aware of the need to provide a better and longer education for the future generations of engineers as we realize how dependent the economic performance of our country and the standard of our environment are upon the quality of our professional engineers. An historical view must find a place in any revision of our courses.

Notes

1. G.W.F. Hegel, *Philosophy of History*, Introduction.
2. S.B. Hamilton, 'Why engineers should study history', *Newcomen Society Transactions*, 25, (1945-7), 1-10.
3. Newton's letter to Robert Hooke, 5 February 1676.
4. See for example M. Claggett, *The science of mechanics in the Middle Ages*, University of Wisconsin Press, 1959.
5. R.J.M. Sutherland, 'The introduction of structural wrought iron', *Newcomen Society Transactions*, 36 (1963-4), 67-84.
6. J. Heyman, 'Beauvais Cathedral', *Newcomen Society Transactions*, 40 (1967-8), 15-35.
7. R. Mainstone, 'The structure of the Church of St. Sophia', *Newcomen Society Transactions*, 38 (1965-6), 23-49 and 'Brunelleschi's Dome of S. Maria del Fiore', *Newcomen Society Transactions*, 42 (1969-70), 107-26.
8. S.H. Adamson, 'Seaside piers', Batsford, 1977.

Sir Proby Cautley (1802-1871), a Pioneer of Indian Irrigation

JOYCE BROWN

The name of Sir Proby Cautley, K.C.B., F.R.S., has not passed unnoticed in historical accounts of irrigation in India, such as they are, and he has been rightly ranked among the pioneers. The network of perennial canals which covers northern India and Pakistan was begun in the 1820s, and has gone on being extended well into this century. For a large part of the period, the work was carried out by military engineers; and even after the creation of a Public Works Department in 1854, it was many years before its senior posts were held by other than military officers. Much of the engineering depended on knowledge gained empirically, and in this respect Cautley belonged to the first generation of canal builders. As an artillery soldier he had received no engineering training, and there was no practical experience in India to guide him. Yet he was responsible for building what was then, and has remained, one of the largest canal systems in the world. There was no model in India to which he could turn for a solution to the particular hydraulic problems of the Ganges Canal, and one can only suppose that he achieved what he did by sheer force of intellect. His influence was felt on several generations of canal builders, while the Ganges Canal itself remained a point of reference for every canal work executed in India in the nineteenth century. It is not too much to claim that Cautley was the father of the perennial canal system in northern India.

Of the man himself, research has revealed some particulars, and a fuller account is given elsewhere.[1] Proby Thomas Cautley was born in the village of Roydon (now known as Raydon) in Suffolk on 3 January 1802. He was the second child born to the Reverend Thomas Cautley and his wife Catherine, née Proby, an older brother born in 1800 having died in infancy. Later, a brother and two sisters were born.

The Cautleys were by origin a northern family. Proby's father (c. 1756-1817) had attended school at Bolton in Yorkshire before entering Trinity College, Cambridge, of which he became a Fellow in 1778. He lived most of his life there, holding various offices of the university, until a late marriage in about 1796; this ended, however, two years later with his wife's death following childbirth. His second marriage, to Catherine Proby, took place at Stratford St. Mary, on 23 December 1799. Proby's grandfather, Thomas Cautley (d. 1796), was also a clergyman, appointed first to the diocese of Chester in 1752 and later to the diocese of York.

The Probys originated in Ireland, and Proby's mother (*c*. 1772-1830) was the second daughter of a large family. Her father, the Reverend Narcissus Charles Proby (*c*. 1737-1804) was rector of the nearby parish of Stratford St. Mary, which Cautley's father combined with his own on his father-in-law's death.

The Cautley family was comfortably off. Despite his father's early death, Cautley received a good education at Charterhouse and Addiscombe Military Seminary, a military school near Croydon set up by the East India Company to train boys for the Indian service. Cautley was a cadet from July 1818 until April 1819, when he was assigned to the artillery and left for India, arriving in September 1819.

Cautley remained in India, with one three-year furlough, until his retirement from the service in May 1854 with the eventual rank of honorary colonel. He was superintendent of the Eastern Jumna Canal from 1831 until 1836, when he became Superintendent General of Canals with authority over the canals of the district; in 1843, he was relieved of executive duties but remained in administrative control of them; in addition, from 1848 until 1854 he was Director of the Ganges Canal works. For his contribution to palaeontology, together with Hugh Falconer he received the Wollaston Medal of the Geographical Society in 1837; he was elected a Fellow of the Royal Society in 1846 in acknowledgement of his work in both palaeontology and hydraulic engineering; he was knighted for his services in India in July 1854; and he was honoured by being invited in 1858 to serve on the first Council of India set up after the Indian Mutiny to rule India in place of the East India Company. He served until 1868, when he retired to Sydenham with his second wife, Julia, née Richards, whom he had married in 1865. His first marriage, to Frances Bacon, in Mussoorie on 20 September 1838 had broken down in 1846 and ended in divorce by Act of Parliament in 1850. The only child of Cautley's first marriage, Walter George, was born on 30 July 1840 and died in 1846. Cautley himself died from bronchitis at Sydenham on 25 January 1871.

The importance of irrigation in India can easily be understood in relation to its climate and physiography. Most rain in India falls during the south-west monsoon, which lasts from July until the end of September. The cold season extends from October until March, succeeded by very high temperatures in April, May and June. The average rainfall for the whole country is about fifty inches per annum, but some areas may receive as little as five inches while others receive five hundred. But the quantity any area receives cannot be relied on even from year to year, for the monsoon may arrive late and withdraw early. This failure of the rains at times when crops need water can be critical, resulting in famine. Areas with less than ten inches of rain per annum require irrigation. Even an area with fifty inches may suffer if there is a failure of, say, a third of the rainfall, since such areas are probably committed to rice-growing, which requires a great deal of water. Such areas may statistically expect ten dry years in fifty, three of them of severe drought. Irrigation is therefore needed in most of India to compensate for the deficiencies of rainfall — its total absence at some times of the year or its periodic inadequacy in due season (Figure 1).[2]

The physiography of India shows broadly two areas: the alluvial Indo-Gangetic plain in the north and the rocky peninsula of the south. The northern area is watered by the great Indus and Ganges rivers and their tributaries.[3] The southern plateau also has large rivers, but flowing through broken country and not easily exploited for water supply. On the eastern belt of the peninsula, the rivers have formed deltas of alluvial sediments. These physical features have dictated different irrigation solutions for each area: an extensive perennial canal system in the north, and widespread use of storage works in the southern peninsula, with deltaic canals on its coastal belt (Figures 2a and 2b).

The climate and physiography, therefore, prompted the practice of irrigation in India from very ancient times.[4] The three early methods of irrigation in a more sophisticated form are still the main components of irrigation in India. Perhaps the most common method depended on wells, often constructed of brick masonry and sunk to great depths to tap water in the sub-soil. A second method utilized storage reservoirs, created by building low embankments on gently sloping ground, the water retained being absorbed or sometimes let through primitive sluices to irrigate land lower down. Artificial lakes or 'tanks' might be created by damming rivers, which in some areas improved the possibilities of well irrigation by raising the water-table. A large number of these tanks, some very ancient in origin, still exist in the Madras area.

A third method was irrigation by canals, of which one can distinguish three types. An ancient method, much employed on the River Indus, was the formation of inundation canals, shallow cuts made through the river bank into which water flowed when the river was in flood. These non-perennial inundation canals relied on elevated river levels for their supply. Perennial canals also took off from large rivers; by means of head works to regulate the supply, they could be made to contain water throughout the year. Many perennial canals are in the Indo-Gangetic plain. A third type of canal was the deltaic canal, in which the water supply was obtained by damming a river near its mouth to divert the water into a fan-like arrangement of channels over the delta. Such was the method adopted on the Rivers Krishna, Cauvery and Godavari, for example.

In addition to these main irrigation methods, water for drinking or watering crops was obtained locally by lifting water from wells, streams and depressions, frequently by simple mechanical means such as the basket scoop, the doon (an oscillating trough), and the lat or picottah, a device known elsewhere as the shadouf. The lat consists of a bucket attached to a pole which can move up or down in a vertical plane; the short end of the pole is counter-balanced by a weight. The operator uses his own weight to depress the bucket into the water, from which it rises

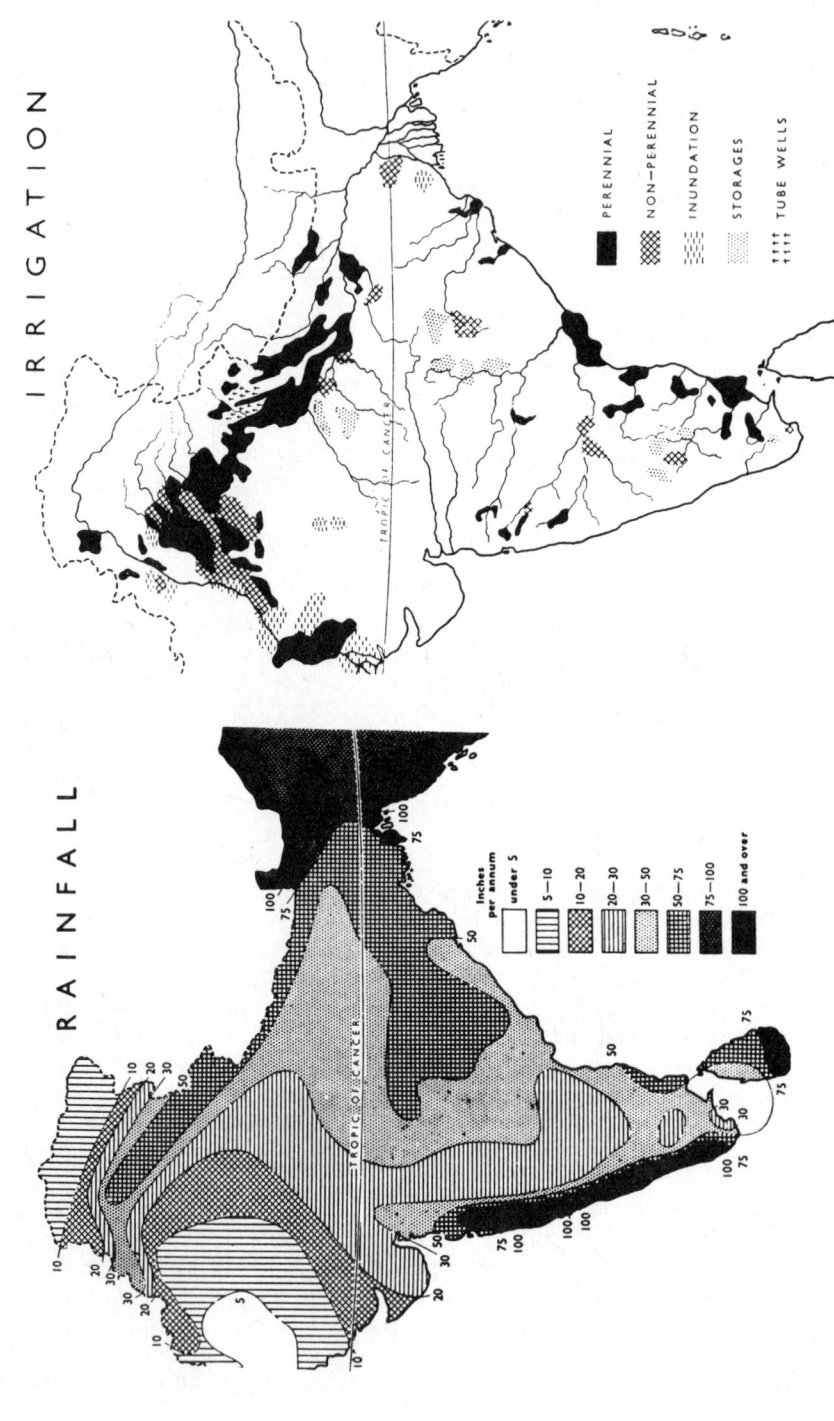

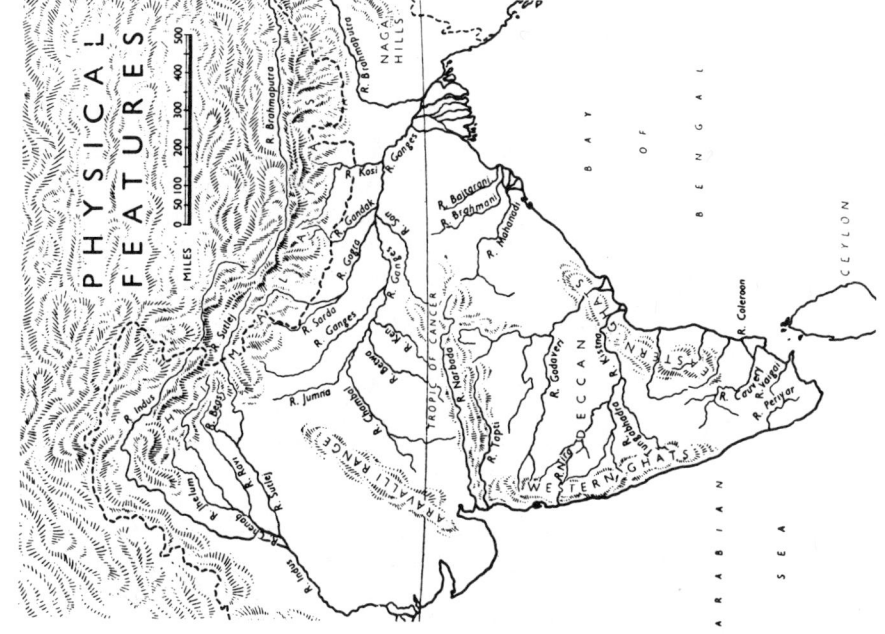

Figure 1. Rainfall of India, 1950 (from F. Newhouse and others, *Irrigation* . . ., 1950, p. 34)
Figure 2a. Irrigated areas of India, 1950 (from F. Newhouse and others, *Irrigation* . . ., 1950, p. 35)
Figure 2b. Physiography of India (from F. Newhouse and others, *Irrigation* . . ., 1950, p. 34)

full by the action of the counter-weight. The lat is suitable for lifts of from four to ten feet. Deeper water can be reached by the Persian wheel and the mote.

In 1903, when the Indian Irrigation Commission submitted a long report on the contemporary state of irrigation in India, it was stated that 293 million acres were cropped annually, of which 53 million (18%) were irrigated: 19 million by canals, 16 million by wells, 10 million by tanks and 8 million by other means.[5] Fifty years earlier nowhere near this acreage was under irrigation; indeed, merely in the two decades before 1900 the area under irrigation increased by a third.[6]

Cautley's work in India was concerned with perennial canals, very few of which had been cut prior to British rule. One notable one, later known as the Hasli Canal, had been built by Sikh rulers to carry water 130 miles from the River Ravi to Lahore. There were also two canals taking off from the east and west banks of the River Jumna and rejoining the river at approximately the same point near Delhi. The Hasli Canal still functioned, but the Jumna canals had fallen into disrepair (Figure 3).

During the three decades when Cautley was in India, the following works of canal irrigation were realized: the reopening of the Jumna canals, the western in 1819 and the eastern in 1830, with improvements being made to them into the 1840s; the cutting of new water-courses in the Dehra Dun in the foothills of the Himalayas between 1838 and 1844; the building of a large new canal, the Ganges Canal, between 1842 and 1854; the incorporating of the Hasli Canal into another new cut, the Bari Doab Canal, made from the River Ravi, between 1850 and 1859; and work in Madras on the delta systems of the Rivers Cauvery and Godavari in the 1830s and 1840s. Cautley was directly associated with the Eastern Jumna Canal, the Dehra Dūn water-courses, and the Ganges Canal; and Robert Napier, who had been one of his assistants on the Eastern Jumna Canal, made the initial survey for the Bari Doab Canal.

To understand why these works were being undertaken at all, it is necessary to recall the position of the East India Company in India. Chartered as a trading company in 1599, the East India Company had established essentially three bases in India, at Bombay, Madras and Calcutta. Britain, however, was not the only European country interested in trade with the east, and the East India Company soon found itself obliged to defend its depots against other powers, notably the French, acting alone or in alliance with native chieftains. The East India Company acquired an army, consisting of Indian troops under British command, and a minority of European troops enlisted at home or from white mercenaries in India. In 1754, the first complete regular battalion of the British army arrived in Madras, and by the end of the eighteenth century there were two or three British battalions.[7] Although the two armies had to work closely together, there was no cross-posting between them. From 1809, the East India Company had its own military

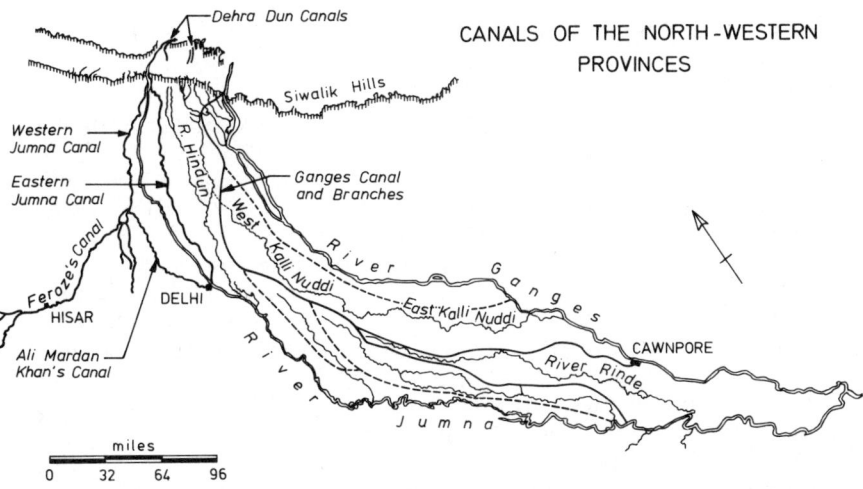

Figure 3. Sketch-map of the canals of the North-Western Provinces, 1854

seminary at Addiscombe, in which it could train gentleman cadets for the infantry, the artillery, and the engineers. Whether it wished it or not, the East India Company found itself obliged to assume the role of a military power.

By 1793, the main struggle was over. The defeat of the French at the battle of Pondicherry left the British as the only European power in India. Two further victories in 1799 and 1803 brought troublesome native states under control.[8] The successful campaigns of the Marquis Wellesley during his governor-generalship (1797-1805) made possible an era of consolidation and a period of British rule not wholly preoccupied with military considerations, but also with the arts of peace. Although campaigns continued to be fought in the first fifty years of the nineteenth century — against the Mahrattas (1817-1819), against Burma (1824-1825 and 1852), against Afghanistan (1832-1842), against Sind (1842), and against the Sikhs (1845-1846 and 1848-1849) — there also began to appear in the East India Company's governance a desire to promote administrative and legislative improvements for the benefit of the native race. Lord William Bentinck (1828-1835) was perhaps the first governor-general conspicuously concerned with improving the lot of the natives. By his efforts, for example, suttee, the practice of burning widows, was abolished in 1829; he introduced Indians into public service offices and tried, by promoting education in the English language and literature, to improve communications between the two races.

The controlling power of the East India Company had, in a series of moves, gradually passed into the hands of the British governmment: in 1773, an act of Parliament established the primacy of Bengal with its governor-general approved by the Crown; in 1784, a Board of Control was set up in London to control the Company in every respect except commerce; and in 1813, the Crown took authority over commerce as well.

In this period of relative peace and consolidation, the government began to turn to the idea of providing public works. From the 1820s onwards, some effort was made in this direction, but the major works belong to the mid-century under Sir Henry Hardinge (1844-1847),[9] and then under Lord Dalhousie (1848-1856).[10]

The principal public works engineered in the period were on a very large scale: a start was made in 1828 on the Grand Trunk Road, which eventually ran a distance of 1,500 miles from Peshawar through the Khyber Pass to Kabul city; a railway was built from Calcutta to Benares, of which the first 120 miles was opened in 1855; telegraphic connections between the main cities were made in the 1850s; and the various irrigation works already mentioned were constructed.

Until 1854, all public works except for railways were executed by army engineers working under the Military Board. They were financed from ordinary expenditure, and no separate capital or revenue accounts were kept. The remunerative character of some of the works, such as canals, however, led to the decision in 1854 to set up a Public Works Department, with a separate system of accounting and staffed by military engineers until late in the century. The history of irrigation in India shows several distinct periods; the period of the first ventures, mainly restoring and improving existing canals (1817-1836), a second period of large new works (1836-1866), and a third period after 1866, the date from which irrigation works were financed by public loans.[11]

The Doab Canal and the Dehra Dūn Watercourses, 1825-1843

Cautley spent his first years in India in military duties, and then early in 1825 he was sent to northern India to assist Captain Robert Smith of the Bengal Engineers, who was restoring one of the old Mogul canals which took off from the River Jumna (Figure 3).[12]

It is not clear why Cautley, an artilleryman, was sent. He had passed through Addiscombe even more rapidly than was usual, spending less than a year there. The seminary had been founded by the East India Company in 1809 to train its own cadets, prior to which time it had paid for a certain number of boys to be trained at Woolwich Military Academy. Only forty-six could be taken there annually, however, and it proved an expensive business. At Addiscombe, sixty boys could be trained, admitted at the preferred age of sixteen. Cadets stayed for four terms (two years) unless they were found on Public Examination to be 'qualified for the Scientific branches of the Profession in less than 4 terms'.[13] The emphasis of the course was on mathematics and fortification, and the scales of merit, up to 1826, are given as: mathematics 28, fortification 28, military drawing and surveying 12, civil drawing 4, Hindustani 12, French 8, Latin 4.[14] The best students were reserved for the engineers, and from 1811 onwards they remained in England, without prejudice to their rank, under Colonel William Mudge for six months on the Trigonometrical Survey.[15] From 1815 onwards, cadets spent an additional twelve months at Chatham under Lieutenant Colonel Pasley, receiving instruction in the practical parts of mining.[16] Cautley was not among those to receive this training. At his Public Examination in April 1819, nine cadets were reserved for the engineers, and the other twenty-nine, including Cautley, were assigned to the artillery.[17] He received a prize for drawing, was commissioned Second Lieutenant on 19 April following, and left immediately for Bengal, being admitted to the service on 11 September 1819 on his arrival.[18] A possible explanation for his hasty passage through Addiscombe was that an urgent request reached the seminary in 1818 asking for artillerymen.[19] Colonel Mudge, in his report of the Public Examination of 6 April 1819, made a plea for candidates in future to be allowed to stay longer.

It seems, therefore, that Cautley arrived at the headquarters of the Eastern Jumna, or Doab, Canal completely ignorant of hydraulic engineering and not much inferior in this respect to the man in charge of the project, Robert Smith. Smith was working as Garrison Officer at Delhi and was showing a certain flair for restoring old Mogul buildings, as well as for making paintings of them.[20] His earlier career had proved Smith to be a competent surveyor, and the government in assigning him to the canal may have thought that surveying was the basic talent that was required. Appointed to the canal on 31 December 1822, Smith completed his survey by the following May, and began clearing out the bed in 1824.

The canal was one of the two mentioned earlier which took off from either side of the River Jumna. Their early history is given in a paper by Major John Colvin, later Superintendent of the Delhi Canal.[21] The canal on the western bank was the older of the two, dating back to the reign of the Emperor Feroze Shah Tughlak in the fourteenth century when it was constructed to carry water from the Jumna to a royal hunting ground at Hissar; it is sometimes called Feroze's canal. It was renovated in about 1568 by the Emperor Akbar, and again about a hundred years later when an extension branching in a south-easterly direction from Mudlouda brought the canal to Delhi. This extension, engineered by Ali Mardan Khan, was known either as Ali Mardan Khan's canal or the Delhi Canal; under the British it and Feroze's canal together became known as the Western Jumna Canal. By 1760, however, this western canal was no longer functioning and needed repair (Figure 3).

It is thought that the canal on the eastern bank was probably cut in the seventeenth century at the time when the western canal was being extended to Delhi. But there was little surviving evidence in the form of old water-courses leading from it to suggest that it had functioned for long. It ran on a steep slope in the north, and was cut at right-angles by four *raos*, or seasonal torrents, for the control of which no provision had been made (Figure 5).

The land on which these canals lay did not come into British possession until 1803. The idea of restoring them was probably insinuated into government thinking when a Mr. Mercer, a private individual, offered to restore the Delhi Canal at his own expense if he could keep any resulting financial benefit for the first twenty years of its operation. The government refused, and sent its own military surveyors to the area. The restoration of the canals had obviously some financial appeal for the government, if not in the return from water rent, at least in improved land revenue. Moreover, the provision of irrigation in areas where well irrigation was expensive because of the depth of water below the sub-soil, or in areas where no wells existed, would facilitate the settlement of waste areas after the recent campaigns.[22]

Surveys were carried out by various officers from 1810, but the matter fell into abeyance until 1817 when Lieutenant George Blane was appointed to restore the 185 miles of Ali Mardan Khan's canal on the western bank.[23] Water was readmitted in 1819 while work was still in progress. In 1820, John Colvin was appointed to restore the 240 miles of Feroze's canal. The canal was reopened in May 1825, and the work of improving it and the Delhi Canal went on throughout the 1830s and the 1840s. That a new kind of military duty had come into being was recognized by the government in its appointment of Blane in 1821 to a newly created post, Superintendent of Canals in the Delhi Territory, but his premature death that year from malaria led to the eventual appointment of John Colvin in 1827. On Colvin's departure from India in 1836, one of his assistants, William Baker,[24] succeeded him on the Delhi Canal, and the latter's report on this canal can be consulted.[25]

Meanwhile, on the eastern bank, Lieutenant Henry Debudé had been sent in July 1822 to survey the old canal.[26] He was relieved of his duties in March 1823 when Captain Robert Smith took over.

The original canal took its head at Kharra and ran south for 134 miles, more or less parallel to the River Jumna, on the high land between the Jumna and the Hindun, curving inwards in a south-westerly direction to flow into the Jumna. It was led from the Jumna along a *nullah*, or stream, to the village of Nyashur, from which an excavated channel brought it to the Raipur *nullah*, or water-course, down which it ran for 2 miles. Taking a south-easterly course over the high land, it passed the drainage of the Jatunwala and Nogong *raos* and entered a ravine at Behut, down which it passed to a junction with the Muskurra river. It was carried along the river for 4,000 feet to Kulsea, from which point excavation brought it on high land to the Shamli *nullah*, along which it was conveyed for 24 miles to Bhynswal; from there excavation brought it south to the Sikrani *nullah*, down which it proceeded to the Jumna. The advantage that the original projector had thus taken of the course of existing rivers and streams meant that in some places the channel was very winding (Figures 4 and

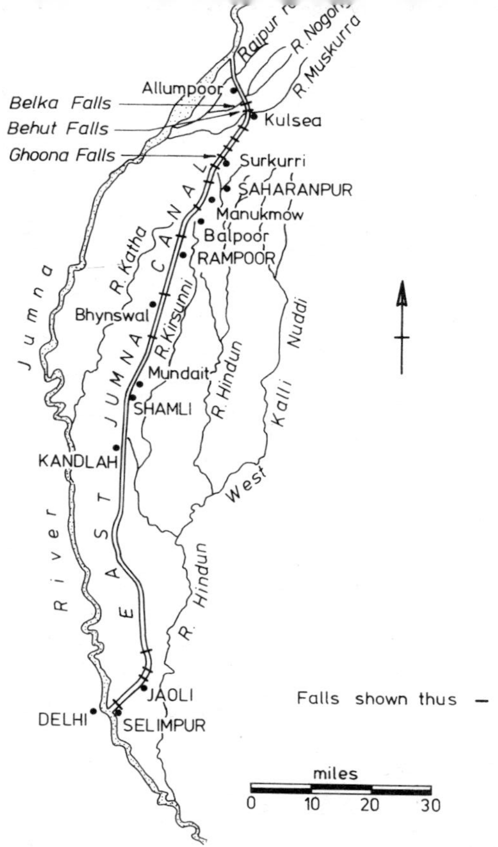

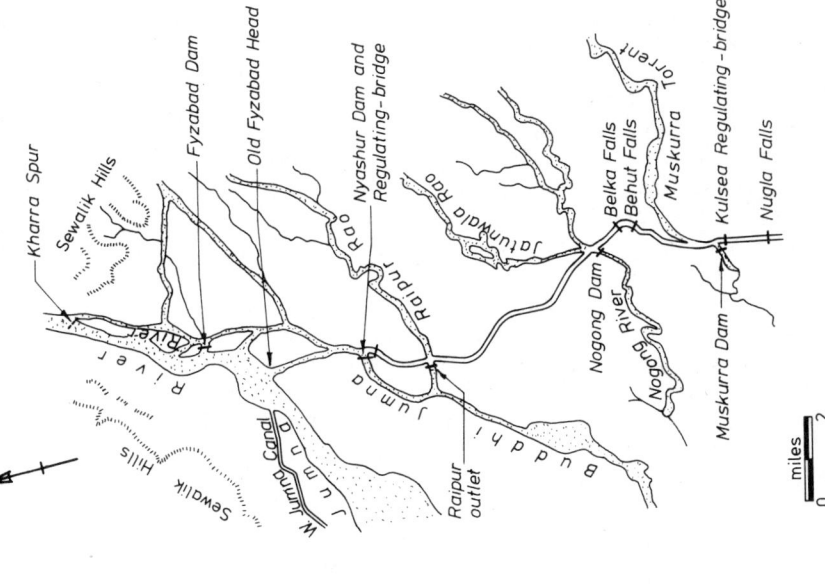

Figure 5. Sketch-map of the head of the Eastern Jumna Canal, 1845

Figure 4. Sketch-map of the Eastern Jumna, or Doab, Canal, 1845

5). Travelling, however, as it did along much of its course on the summit of the ridge separating the Hindun and Jumna rivers, it was ideally suited to water the gently sloping land on either side. Cautley's report on the canal provides details for the account that follows.[27]

Smith's line was much the same as the original one except that he planned to establish the head a little lower at Fyzabad, and instead of letting the tail-water escape down the Sikrani *nullah* to the Jumna, to make a cut from Gokulpoor to Selimpoor. The most tortuous stretches of the canal were to be straightened out.

Excavation of the channel had been started in different places by the time Cautley arrived in 1825. No masonry works had been constructed, but a start had been made on a scheme to deal with the seasonal flow of the worst of the mountain torrents, the Muskurra River. Three cuts had been made from it and an embankment built across it to try to deflect its flood waters into the tributaries of the River Hindun. One of the first tasks Cautley was given in May 1825 was to clear out the mouths of these cuts. Regulating works and other diversionary techniques were employed to deal with the other cross-drainage problems created by the seasonal torrents of the Raipur, the Nogong, and the Jatunwala. An inlet and outlet were made in the canal bank to pass off the waters of the Raipur, while the waters of the Jatunwala were deflected into the Nogong river and controlled by a masonry dam across its bed.

The next five years were spent in clearing out the channel, straightening out the worst bends, and beginning the construction of masonry works. A sketch book, almost certainly Cautley's, has survived, containing beautifully executed hand-coloured drawings of some of the bridges on the canal, elegant because they are drawn on such a small scale.[28] The drawings show elevation, plan, and section for bridges of several designs and dimensions, and are sometimes dated with dates between 1828 and 1833. Occasionally the name of the builder is given. There were several large two- or three-arched bridges, but the majority were single arch of between 15 and 20 foot span. Twenty-three are of this type; another twenty-six have a single arch with a circular opening on each side of it of 4, 5 or 6 feet in diameter. The book shows designs for inlets and outlets, and one or two elaborate structures where a bridge, escape, and ghat, or bathing steps, formed one unit. The foundations of the bridges were usually of the type Robert Smith had designed, laid on shingle or boulder, with curtain walls to the front and rear built to a depth of 6 feet, their waterways counter-arched with brick masonry.[29]

The dams on the Muskurra and Nogong rivers and one at Nyashur across the Boodhi Jumna were all built during Cautley's first years on the canal. The Muskurra dam at Kulsea is typical of the design employed. It had eighteen openings, the six centre ones fitted with gates, the others controlled by sleeper planks, known elsewhere as stop

---

Figure 6. Cautley's drawing of the design of well-foundations (published in *Journal of the Asiatic Society of Bengal*, Vol. 8, 1839, Plate 2)

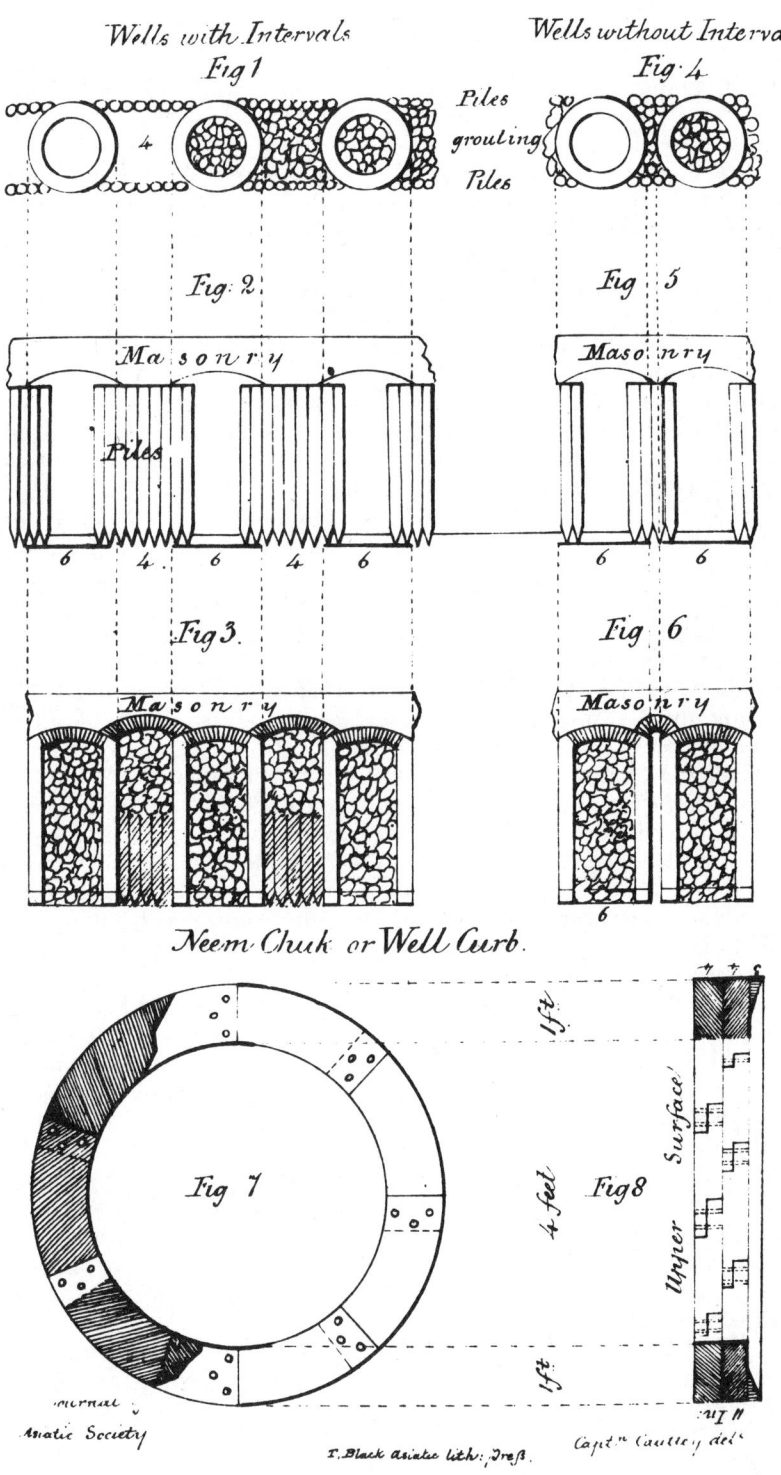

logs. Its foundations were made as strong as possible by the use of wells constructed of brickwork, placed close together and sunk by the native method of well-sinking, or an adaptation of it. The wells were filled with earth while the space between them was packed with piles and sometimes concrete, and then covered by an arch of brick masonry with an additional arch thrown from one well to another (Figure 6). Cautley described this type of foundation in a paper to *Journal of the Asiatic Society of Bengal* published in 1839.[30] The concrete was made of *kunkur* (naturally occurring alluvial lime rock), broken river stone, and the *gutta*, or refuse, of lime kilns, mixed with a cement made up of two or three parts of *soorkhee*, or pounded brick, and one part of best stone lime; this employment of concrete in foundations must be a very early use in India. For the Muskurra dam, three parallel lines of wells were constructed, the front and rear lines sunk to a depth of 12 feet, and the centre to a depth of 6 feet. The wells in the front and rear were placed 6 or 8 feet apart, the space between being filled by a masonry box undersunk to 6 feet. Lines of 16 foot piling were driven at a distance of 10 feet from the body of the dam to the front and rear, and the intermediate spaces filled with broken brick, clay and refuse material.[31] Despite these strengthening measures, Cautley recounts that the Muskurra dam was damaged in floods in 1829, 1831, 1835, 1842, and 1844, which gives an idea of the force of the waters with which he had to contend. The Nogong and Nyashur dams also suffered in flooding, and the Nogong dam had eventually to be rebuilt in 1842.

The dams at Kulsea and Nyashur were accompanied by regulating-bridges on the canal, so that, working in conjunction with one another, the volume of water passing into the canal could be controlled. At Kulsea, the bridge had three spaces, the centre one of 20 feet and the two side ones of 15 feet. The openings were fitted with gates and sleepers.

The bed of the canal ran on shingle or stone boulders from its head to Allumpoor, but from there to Surkurri it ran mainly over sand. From Surkurri to Jaoli, the bed was of clay, but from Jaoli to its tail at Selimpur, there was again a long stretch of sand. The sandy beds also coincided with the steepest slopes of the canal. There were two main steps, with a fall of 186 feet in the first 28 miles and a fall of 45 feet in the last 11 miles, with a fall of 189 feet in the intermediate distance of 94 miles (Figure 4).

By the end of 1829, work was sufficiently far advanced for Smith to contemplate the admission of water to the canal works. He was in poor health and anxious to return to Europe, so was not present when the canal was officially opened in the presence of Captain Colvin on 3 January 1830. No-one was appointed immediately in his place, and the task of administering the canal works in the first months after the opening devolved upon Cautley.

Immediately problems arose. It was soon apparent that the two fundamental problems posed by the canal, the slope and the stretches of sandy bed, had not been tackled. The water rushed unimpeded down the slope carrying with it large bodies of silt which were deposited in stretches where the current was slow. Rapids formed at different points on the steep slopes between the bridges, and working back, they began to expose the foundations. By the twentieth of the month, the Belka and Ghoonna bridges in the north and the Jaoli and Selimpur bridges in the south were under severe strain.[32]

All Cautley's energy was taken up in trying to deal with these problems. With the assistance of his overseers and native help, he built rafts of timber to place in front of the Ghoonna and Belka Bridges in an effort to save them, but he could not prevent the Belka Bridge from being washed away. He and his officers ranged the banks building up the embankments to try to keep the water in the channel. Speaking of this time afterwards, Cautley wrote, 'I can only recur to the anxiety and care of the period when these works were under construction — to the difficulties that were experienced in contending against these rapids, and in maintaining the supply of water in the Canal, under the deposits of silt hourly forming in the central regions, requiring constant raising of embankments on extended lines . . .'[33]

The next twelve years of Cautley's life were devoted to the canal. He was officially appointed Superintendent of the Doab Canal in April 1831, and, following Colvin's departure at the end of 1836, Superintendent General of Canals with authority over the canals in the district. This title, however, was shortly afterwards abolished, and instead he became Superintendent of the Eastern Jumna Canal, but with the same duties.[34] In 1843, he was promoted to purely administrative control of these canals, and Richard Baird Smith succeeded him as Superintendent on the Eastern Jumna.[35] Smith was later also to succeed him in his title Superintendent of Canals of the North-West Provinces and Director of the Ganges Canal Works.

Cautley had immediately to devise a project to check the destructive and erosive velocity of the stream. Work was promptly begun on the building of falls at the worst places, and during the next four years five were built in the northern step and four in the southern, all about 2 miles apart.[36] In the northern step, the descents were one of 15 feet, two of 8 feet and two of 7½ feet, while in the south there were two of 8 feet, one of 7½ feet, and one of 15 feet at Selimpur at the tail connected with four corn mills. They were designed in three chambers: a central one 20 feet wide and one on each side of it 15 feet wide to allow one of the chambers to be developed later for lockage.[37] The water descended over an ogee curve, a design intended to deliver the water on the lower level as gently as possible (Figure 7). Since, as far as we know, there were no models Cautley could copy in India, these may have been the first falls introduced in the country.[38] Foundations were made as solid as possible, using wells 6 feet in diameter sunk to a depth of 10 or 13 feet (Figure 8). Where possible, mills for corn-grinding were attached.

The action of the falls in controlling silt deposition proved favourable, for although much silt was carried forward, it was laid down very

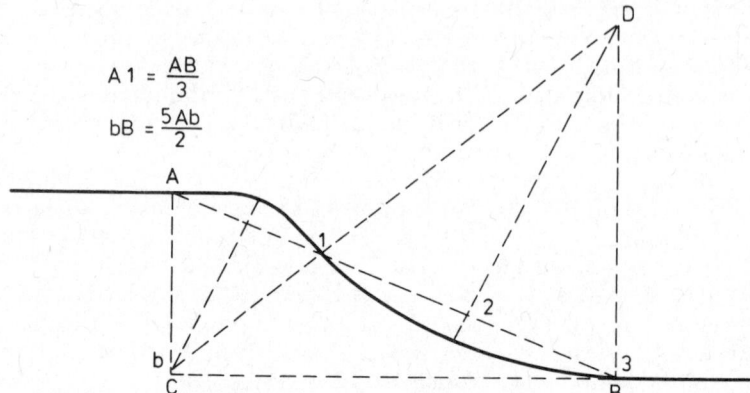

Figure 7. Design of an ogee fall (redrawn from Cautley, *Report on the Ganges Canal Works* ..., 1860, Vol. 1, p. 157)

gradually and Cautley was able to keep the embankments up to the encroaching evil. In reviewing the situation in 1836, Cautley could see that what seemed to be successful was a slope of 1 inch per 100 yards, or 17.6 inches per mile in a hard soil, with the maximum desirable slope 24 inches per mile. In his *Report on the Levels of the Doab Canal* addressed to the Secretary of the Military Board and dated 1 April 1837, Cautley proposed a slope of 17.6 inches per mile for the northern section, which made necessary a new 7 foot descent between the Belka Falls and the Muskurra River.[39] Below Surkurri, where the soil was of *kunkur* and clay, he proposed a slope of 24 inches per mile, to be achieved by building a series of nine falls with descents of 4 feet each. Below the last of these, the proposed slope was to be 20 inches per mile, an effect to be obtained in part by lowering the floorings of four bridges on this part of the line.[40] With modern knowledge we can say that all these falls had the effect of decreasing the velocity of flow and thus controlling the worst effects of scour. Cautley's theory was that the influx of sand and silt would not equal the amount removed by the current and that the large deposits would gradually be demolished and moved forward. Although this view was a controversial one when he put it forward, it proved correct and the larger deposits disappeared.[41] By 1845 when Cautley submitted his final report on the canal, the additional fall in the northern step and five of the small ones proposed had been built, while the four bridges had been demolished and replaced. Even with less silt deposited, however, it meant over the years massive raising of embankments, particularly in the central division where the canal was carried in an earthen channel sometimes as much as 10 or 12 feet above the surface of the country for lengths of 30 or 40 miles. The risk of water escaping through burst embankments was thus constantly present and required steady vigilance by the overseers and their staffs. Silt also affected the headroom at bridges, many of which had to be remodelled later.

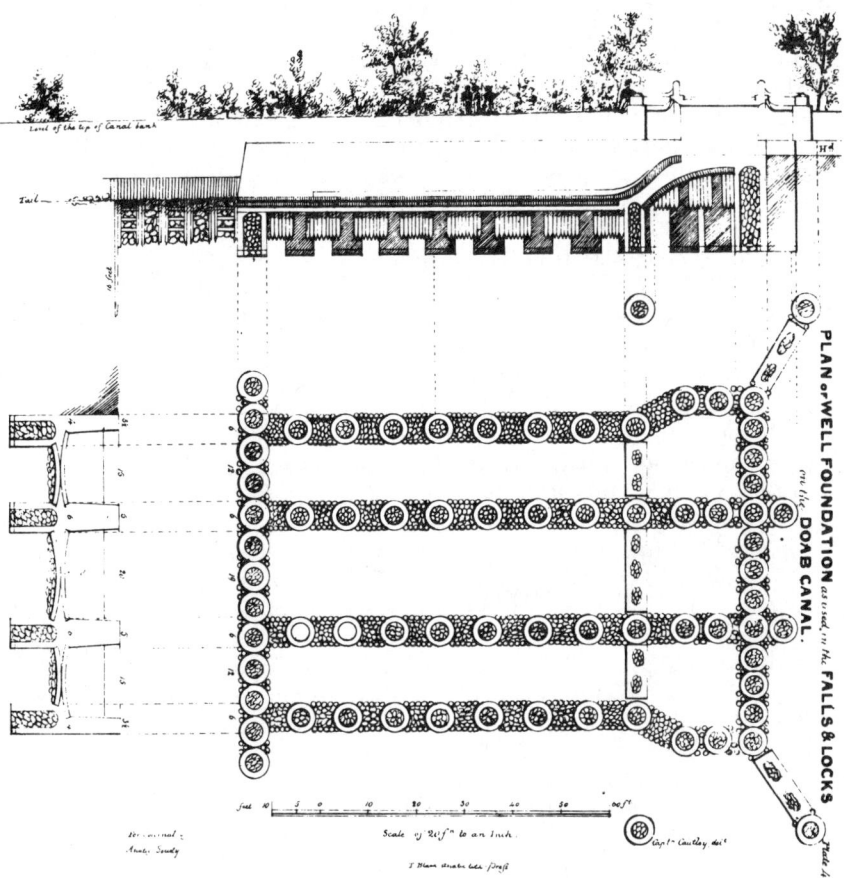

Figure 8. Cautley's drawing of the plan of well-foundations as used on the falls and locks of the Doab Canal (published in *Journal of the Asiatic Society of Bengal*, Vol. 8, 1839, Plate 4)

Work went on steadily over the years with the building of the falls, mills, bridges, water-courses and the carrying out of repairs to the dams when heavy flooding had damaged them. In clearing out the canal bed south of the Belka falls, Cautley was interested to find the vestiges of an ancient Hindu town 17 feet below the ground surface, and submitted a paper about it to the Asiatic Society of Bengal in Calcutta.[42] The bed of the river at the Fyzabad head had retrograded to such an extent that it was decided in 1834 to derive the supply from the Kharra Head (Figure 5) which was done by building a temporary spur out into the Jumna to direct the water at times of low river levels down the old Buddhi Jumna river. The idea was that the spur would be swept away in the first floods every year when the supply was ample and it was desirable to return a large volume of it to the Jumna. A bund was built at Fyzabad across the channel to help in this redirection of the water, and in 1841-43 the bund was replaced by a masonry dam. Bunds helped to ensure the supply for the Delhi Canal, while the dam and regulating-bridge at Nyashur could be used in conjunction with each other to regulate the flow for the Doab Canal. Temporary works of this kind, made of wooden structures or bundles of twigs and filled with boulders, sometimes still are used on Indian canals; the origins of their use is not known.[43]

The canal endlessly tested both the patience and the ingenuity of the canal officers in the face of the annual rise of the river, which might bring massive volumes of water and floating trees and logs to bear on the canal structures. A new kind of self-regulating sluice-gate was introduced in the Nogong and Muskurra dams, which dropped to a horizontal position under the pressure of floodwater and provided a wide opening.[44] To protect the dams from retrogression, Cautley devised what he called retaining dams, built some distance from the tail of the masonry dams and composed of two parallel lines of 16 foot piles driven 40 feet apart across the bed of the river, the space between being filled with frame boxes containing stone, sunk 4 to 5 feet into the river bed and tied together by sleepers.[45]

Cautley was anxious that the canal should provide service as soon as possible. In the first two years, mills, usually attached to the falls, were built in ten places. The water-distribution system was also being developed as rapidly as possible with *rajbuhas*, or water-courses, cut from the canal. Originally, the government had granted 10,000 rupees per annum to Cautley to allocate in small payments to local landowners to enable them to build water-courses from the canal. This, however, turned out to be an inefficient method, since the engineering was not properly carried out and water was often wasted. Cautley therefore had all the *rajbuhas* constructed by canal officers and redeemed the cost from the landowners.[46] There were basically two main lines of water-courses parallel to the canal on either bank with branch water-courses leading from them every four miles of canal length (see Figure 21). By 1847, there were forty-seven main distributory channels, totalling 465 miles.[47] The *rajbuha* heads were either of masonry or earth, and in 1844 it was reckoned there were 270 on the left bank and 322 on the right, of which altogether 113 were of masonry.[48] Native officers assessed the amount of revenue due from the land irrigated, which was collected by four native agents, one in each of four districts. Revenue also accrued from rent of mills, rafting of timber, sale of produce from the plantations built on the strip of land in the rear of the embankments, and from fines imposed for breach of canal regulations, such as establishing unofficial outlets. From 1837-38 onwards, the canal showed a surplus of revenue over expenditure, although there were occasional bad years when there was a deficit.[49] The government, whatever beneficent motives it may have had, was undoubtedly influenced in its decision to promote further irrigation schemes by the profitability that the early canals quickly showed.

The work was organized in three divisions, a northern, a central and a southern. The staff comprised an Assistant Engineer, two overseers and two assistant overseers with a native establishment permanently employed to keep accounts, to operate the gates at dams, escape heads, and regulating-bridges, to look after water-course heads and the state of the embankment, and to collect revenue. Excavation of the channel, which was 6 or 7 feet deep, was by locally recruited native labour and contractors, who were also responsible for consolidating the embankments. A note in Cautley's Doab Canal Sketch Book, where he gives an estimated cost of building a bridge, states that the price will be 25 per cent lower if convicts are used.[50] The overseer sent to the Executive Engineer a weekly report of the progress of works and a monthly account of cash received and expended, showing the number of labourers employed daily, expenditure on materials, and payments to contractors. The Executive Engineer was answerable to the Military Board.

Cautley's Assistant Executive Engineer for five years was Robert Napier, appointed in April 1831, the same Napier who was to become one of the greatest soldiers of his age.[51] Cautley gave him charge of the central division, believing that 'he who has done the duty of an Overseer himself, is best able to superintend and appreciate the labours of an Overseer.'[52]

Cautley's abstract of works constructed under his superintendence shows that most of it was supervised by Sergeants J. Pigott and Petrie in the north and Sergeant H.B. Brew in the south.[53] This staff he considered inadequate in number and he recommended that there should be at least three overseers and three assistant overseers. Bearing in mind the hardship of the life, Cautley was angry when the salary of Assistant Executive canal officers was reduced from 250 rupees to 200 rupees per month.[54] He describes the hard and dedicated life of the canal officer in the wet season, 'the necessity frequently of being exposed during the day in heavy rain, or having to visit the bunds and dams at all periods of the year — to oppose by his own energy, skill and quickness in resources, an element at any time the most difficult to contend with, but in the case of the mountain torrents opposed to his numerous and difficultly situated works requiring the utmost steadiness and decision of character.'[55] The life required men 'who can without grumbling and without being dissatisfied reside in the district, away from society, and separated altogether from the enjoyments of social existence.'[56]

The Doab Canal passed from Cautley's executive command in 1843, but its design problems did not. By 1849 the tortuous section in the north was giving drainage problems, with stagnant swamps developing along the line of the canal between Balpoor and Bhynswal. As Superintendent of Canals of the North-Western Provinces, Cautley requested a survey of this section with a view to remodelling these 35 miles. The resulting suggestions were urged upon the Government in 1850 in a report written by the canal's Superintendent, who took the opportunity to point out that since the government gained so much financially from this profitable canal, it was reasonable to hope for expenditure to be made on protecting the local populace from disease.[57]

The Doab Canal, which still functions today, was capable of irrigating in 1847 an area of 421,875 acres for the benefit of a population of 291,000. The return to government, if one includes the improved land revenue, was estimated to be £19,500 that year on a capital outlay to date of £81,460, or nearly 24 per cent. Although in the bad famine of 1837-38 the canal played only a small part in preventing starvation, the inhabitants did not suffer the misery and mortality experienced by the inhabitants of the Ganges plain.[58] From the 1860s, the Doab Canal proved to be a very profitable work.[59]

While working on the Doab Canal, Cautley was also responsible for the design of three water-courses in the Dehra Dun, a tract of land 48 miles wide from west to east, and between 10 and 15 miles from north to south, lying in the foothills of the Himalayas. They are described in his report of them.[60] These water-courses were comparatively short lengths of channel less than 20 miles long, with widths of less than 5 feet, but they supplied water that was badly needed and which greatly improved the fertility of the area they watered. At the end of 1837, Cautley was sent by the government to draw up plans and estimates for a water-course to be cut from the village of Beejapur to irrigate a triangular tract of land 7,500 acres in extent, west of the town of Dehra. Although only just over 11 miles long, the Beejapur water-course passed through some difficult territory on an enormous slope. Baird Smith describes it:

> The channel, after leaving the Tonse, is carried boldly along the faces of the cliffs forming the sides of the ravine in which the river flows; and, sometimes by cutting through the rocks, sometimes by raising the foundations from the bottom of the ravine, by tunnels in some places, by aqueducts in others, it is brought through difficult ground to the high land at Dhakra, whence it proceeds to Gurki, and is there divided into two branches, one to the eastward, the other to the westward.[61]

The first mile or so was in a masonry channel 5 feet wide and 3 feet deep, and the remaining 8 or 9 miles in earthen excavation. The slope of the country over which it passed was enormous and had to be overcome in no less than ninety-six falls varying from 2 to 8 feet in height. Designed by Cautley in 1837 and 1838, the work was carried out by Captain Henry Kirke between October 1839 and January 1841[62] (Figure 3).

Meanwhile in 1840 Cautley had carried out a survey for the restoration of an old canal called the Rajpur aqueduct, which had since time immemorial supplied the people of Dehra with drinking water drawn from the head of the Raspunnah torrent at the foot of the Himalayas. An undated map of it signed by Cautley has been preserved.[63] The water-course was 12 miles long, carried in a masonry channel for 7 or 8 of them, passing through the town of Dehra and opening into reservoirs at short intervals apart. A branch of it irrigated a triangular tract of land to the east of the town. There were corn-mills on both the Beejapur and Rajpur water-courses. Captain Kirke also built the Rajpur aqueduct between 1841 and 1844.

Cautley designed a third water-course, the Kuttha Puthur, and submitted estimates in 1841, but the project, which he described in the *Journal of the Asiatic Society of Bengal*, was abandoned.[64] It was built a few years later.

Cautley gained much valuable experience from this period of his life. As Robert Smith's assistant, he had the first-hand experience of supervising the building, to Smith's design, of some of the structures. Later, he had the task of modifying them to suit changing circumstances and of designing protective works for them. The falls he introduced were built to his own design, and may have been the first falls seen in India. He extended and developed the practice of using wells as foundations for works executed in sand or loose soil. He learnt a good deal about silt deposition, and investigated the desired slope of channel bed needed to control it. He became all too familiar with the capricious nature and frequent violence of the seasonal torrents in this part of India. But in relation to his principal hydraulic venture, the Ganges Canal, this knowledge was of limited usefulness. The terrain and cross-section of that canal required him to devise elaborate expedients, and its major faults, its slope and the design of its falls, are attributable to the fact that the solutions found to these problems on the Doab Canal were not transferable to a much larger canal, as time revealed.

The Ganges Canal, 1842-1854

PRELIMINARIES

The work of British engineers on the Jumna Canals was of the nature of restoration, in the sense that they made use of the original concept and followed to a large extent the original alignment. Nevertheless, the numerous improvements and novel structures they introduced invite a substantial claim to originality. The Ganges Canal, however, was a purely British work and was the first engineering of its kind undertaken by the British in India. In addition, it was certainly at that date the largest canal ever attempted in the world. There was nothing comparable in scale in Europe. While some very long mileages of channel had been cut in North America — the Erie Canal of 363 miles opened in 1825, and the Pennsylvania Canal of 394 miles opened in 1834 — they were built solely for navigation. Therefore, the Ganges Canal, as a dual-purpose canal, was not only the longest navigable canal in the world, it was also the longest irrigating canal. That such a large canal should

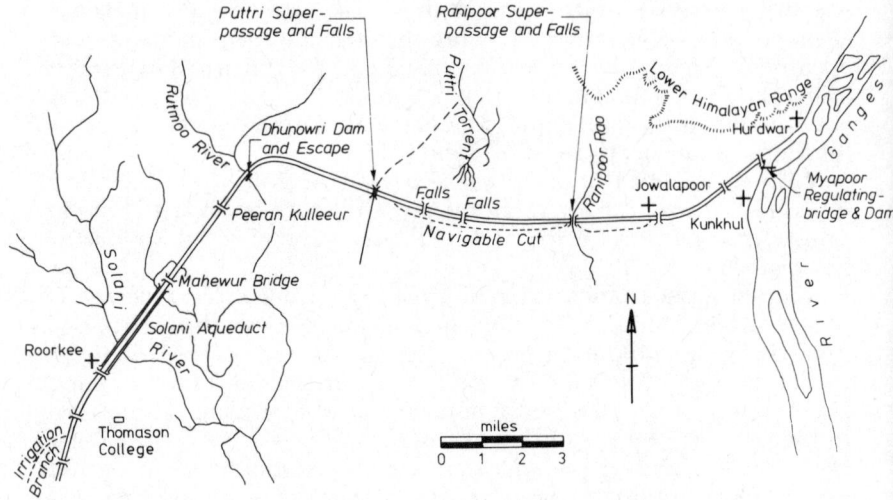

Figure 9a. Sketch-map showing Northern Division of the Ganges Canal, 1854

have been built so early in the history of canal engineering says much for the courage of its projector.

The main sources of information about the Ganges Canal are Cautley's own reports of it and in particular his Report of 1860, published in three volumes with a folio atlas containing engineering drawings of all the main structures.[65] The other reports are the 1840 Report on feasibility,[66] the 1845 Report suggesting three possible projects,[67] and the 1850 Estimate which embodies the scheme which was finally built.[68]

Before John Colvin's departure from the Western Jumna Canal in 1836, its financial success and the protection such works gave in time of famine had aroused government interest in irrigation, and had led him to consider whether a canal should be led off the Ganges to water the land lying between the Jumna and the Ganges. An earlier proposal for a canal in this plain had been made by Captain Debudé, the man who had made the early surveys of the Eastern Jumna Canal. His scheme, however, to dam two of the natural water-courses in the plain on the site of a disused eighteenth century canal about 11 miles long, known as Muhumud Aboo Khan's canal, was soon seen to be impracticable, as the water-courses would not be able to maintain a steady supply.[69] Unlike a purely navigable canal, a canal of irrigation requires a certain velocity of flow in order to convey water to all parts of its distribution system. Colvin's attention turned to using the Ganges since, debouching from the mountains on a steep slope, it would provide the necessary head of water. The problem was to lead the supply across ground running on a slope through rough country to a point from which it could be carried south on a ridge of high land and could

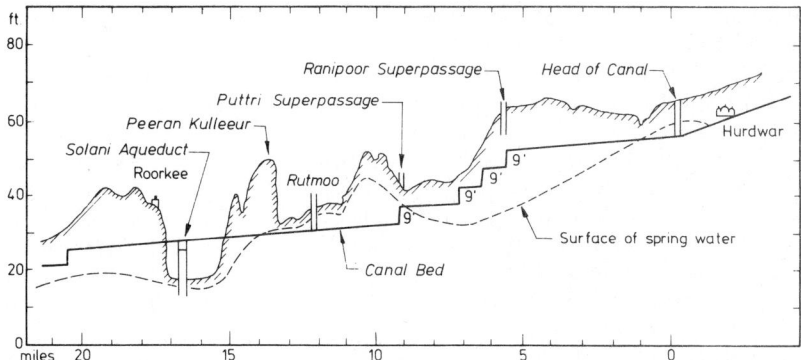

Figure 9b. Cross-section of the Ganges Canal, Hurdwar to Roorkee, 1854 (redrawn from J.G. Medley, *Irrigation Works*, 1873, Plate II)

thus irrigate the plain on either side. The profile of the land in question can be seen in Figure 9b.

In November 1836, shortly after Colvin's departure but encouraged by him in correspondence, Cautley went to the Hurdwar district and took the first series of levels. He concluded that the terrain was very difficult and the matter was left in abeyance. Bad famine in 1837-38, however, brought such general misery to the people and financial loss to the government that Lord Auckland as Governor-General provided a few thousand rupees for a more extensive examination of the *khadir*, or low ground, to the west of the Ganges River.[70]

Early in December 1839, Cautley again went to Hurdwar to reconsider the problem. The Ganges *khadir* is a triangular tract of land bounded on the north-east by the Sewalik hills, on the south-east and south by a steppe or bank, and on the south and east by the Ganges River. East of it, the drainage of the hills collects in three distinct basins. In the wet season, these basins disgorge their contents in seasonal torrents, of which there were four likely to cross the path of any canal led from the Ganges near Hurdwar. In short, it was a similar problem, but on a much larger scale, to the one which had affected the head of the Doab Canal (Figure 9a).

Cautley's report to government, dated 30 June 1840, showed that in essentials the problem could be overcome, and the *khadir* crossed at a cost of 1,000,113 Company's rupees (approximately £100,000 in the values of the day).[71] Much of this sum was accounted for in his daring proposal to cross the Solani River, the worst of the seasonal torrents, in a massive aqueduct nearly three miles long. In a supplementary report of 15 August 1840, he gave the cost of providing 255 miles of main canal and 73 miles of branch, with facility to make the line navigable.[72] The total cost was given as 2,591,158 rupees (roughly £260,000) and he estimated a return of 10 per cent on capital outlay. Whether the government wished the canal to be longer than this was a matter for its own consideration.

The project now entered a period of seven years' discussion and negotiation before the government gave its full-hearted support. Lord Auckland supported the project and set up a committee in November 1841 to supply a more detailed report. This committee was composed of Major Frederick Abbott of the Engineers, at that time Superintending Engineer of the North-Western Provinces,[73] Cautley's friend Captain William Baker,[74] and Cautley himself. Their report, dated 7 February 1842, examined the probable effects of the abstraction of the maximum supply of water on the navigability of the Ganges, about which some anxiety had been expressed; the cost involved in extending the canal to Allahabad; and the estimated financial return.[75] At Hurdwar 8,000 cubic feet per second of water were obtainable, and the committee estimated that the abstraction of 6,750 cubic feet would be sufficient to carry a main line of navigable canal from Kunkul to Cawnpore and to irrigate the area lying between the Ganges on the one hand and the Hindun and Jumna on the other. The government could choose between a scheme incorporating the Solani aqueduct, costing 6,658,848 rupees (£665,884), or one following a more circuitous line in the Ganges *khadir*, without aqueduct, costing 5,936,995 rupees (£593,699). The area irrigated would be 2,303 square miles. A staff under Cautley of three executive officers, six assistant engineers and twelve overseers was recommended.

The scheme had the support of Mr. T.C. Robertson, Lieutenant-Governor of the North-West Provinces,[76] but at this point Lord Auckland was replaced as Governor-General by Lord Ellenborough, who arrived in India in February 1842. War against Afghanistan was still exerting a financial pressure on the government, and Lord Ellenborough himself did not approve of the scheme. The government refused to appoint assistant engineers, but the Agra government nevertheless gave orders on 25 February 1842 for work to begin on the lines of the committee's recommendations.

The first spade broke ground near Kunkul on 16 April 1842, and Cautley, with one assistant whom he moved from the Eastern Jumna Canal, and an uncovenanted assistant from the Canal Department employed to make bricks, began work. In June, however, Lord Ellenborough ordered suspension of the work and the discontinuation of all further expenditure. A few days later he modified his views, and Cautley felt encouraged to write and ask for a modest lakh or two of rupees (one lakh = 100,000 rupees (£10,000)) to enable him to go on working. As he had by July 1842 already spent 138,198 rupees (£13,800), this was a very moderate request.[77] Lord Ellenborough's response was the begrudging grant on 17 September of two lakhs per annum and his proviso that the canal 'should be in the first instance a canal of navigation, and all the water not required for that purpose may be distributed for the purposes of irrigation'.[78] With the interests of

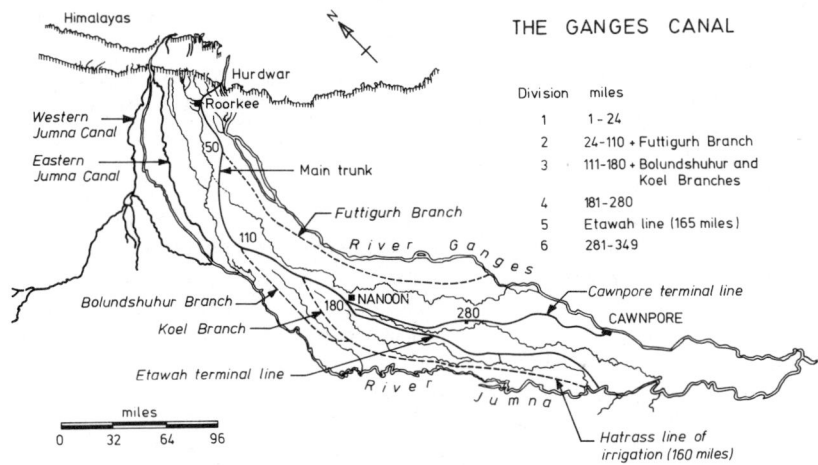

Figure 10. Sketch-map of the Ganges Canal, 1854

navigation in mind, he wanted the canal brought to Allahabad as a terminal rather than Cawnpore, since the steamers ascended the Ganges to Allahabad (Figure 10).

This view of the canal was completely contrary to Cautley's or Colvin's intentions and, indeed, was rather foolish from a technical point of view, since a canal running with some velocity was not going to be ideal for vessels travelling up it. To make the canal dual-purpose added to the difficulties of the engineering, and while Cautley had concurred with the Committee's report of 1842 to provide a navigable line, it had been conceived then as a function of secondary importance.

The work went on in spite of these difficulties and Cautley's position was made easier the following year when in July 1843, he was relieved of his executive duties in connection with the Doab and Dehra Dun canals while remaining in independent control of them. He was also aided by the appointment of an executive engineer for the northern division, who, interestingly enough, was Lieutenant Richard Strachey.[79] Then aged about twenty-six, Strachey later achieved high administrative office in Indian affairs and became the father of the writer Lytton Strachey.

In the cold season of 1843-44, Cautley spent six months carrying out surveys of the Cawnpore and Allahabad districts, doing much of the work himself. The very detailed maps he made appear in the atlas accompanying the Report in 1860.[80] Mr. James Thomason,[81] who had become Lietuenant-Governor of the North-West Provinces in 1843, protested in a letter of 10 February 1844 to the government against this misuse of the Director's time and talents.[82] After a visit to the works, he pleaded the issue to Lord Ellenborough in a letter 11 April 1844, asking if Lord Ellenborough's financial restraint could be justified either on the grounds of policy, economy, or humanity. He pointed out that a supply of artisans had been procured from the thousands of pilgrims assembled for the Hurdwar fair, and the government's prestige would be shaken if the work did not proceed.[83] Lord Ellenborough granted one more lakh per annum. At this point, in July 1844, he was succeeded as Governor-General by Sir Henry Hardinge.[84]

The report of the surveys to Cawnpore and Allahabad was dated 12 February 1845, and offered the government a choice of three projects for the last section of the canal.[85] In two of the schemes, the main line would continue to Allahabad, and in the third, to Cawnpore. The last scheme was favoured by Cautley, and was also the cheapest at 9,339,746 Company's rupees (£933,975). Three or four main branches for irrigation only would leave the main trunk, giving a total length of 878½ miles. Cautley's views on navigation as a primary object were stated firmly, 'my ideas on this subject are unaltered; I consider that all the available water in the Doab properly belongs to the agriculturist and the soil.'[86]

At this juncture with his report in the hands of the government, Cautley, whose health had been failing, decided to take his first furlough in twenty-six years, twenty of which had been spent in what he calls 'the exposed and harassing duties of Canal employment.'[87] Even without fervent government support, certain progress had been made on the canal. Bricks had been made, and collected at numerous points on the line between Hurdwar and a point below Meerut; masonry work had started on a small scale on the dam and regulator at the headworks at Myapoor; channel excavation had begun at several places in the northern division between Hurdwar and Roorkee; workshops had been built at Roorkee and *choki* posts at Myapoor, Roorkee and Munglour; and the route of the canal had been agreed for a hundred miles south of Munglour. This had been achieved by the efforts of Cautley with a staff of two executive officers and two unconvenanted assistants. In February 1845, Cautley left India for England.

His place was taken by Major William Baker, who, it will be recalled, had arrived in India in 1826 and had worked as an assistant to Colvin on the Delhi Canal, succeeding him in the management of the canal in 1836. He and Cautley were close friends, for it was with Baker and Henry Durand, another of Colvin's assistants, that Cautley shared his interest in collecting and classifying fossils.[88] An account of this part of Cautley's life is given elsewhere.[89]

Cautley used some part of his furlough to improve his knowledge of hydraulics.[90] He is said to have visited the Caledonian Canal,[91] built by Thomas Telford between 1803 and 1822, a great engineering feat and one of the show-pieces of hydraulic engineering. In August 1847 he set off to make his way slowly back to India via Italy and Egypt. At this period, the Italians were considered to be leaders in hydraulic engineering, and Cautley was anxious to learn what he could. He particularly hoped to see techniques for dealing with river torrents which cut canals at right angles, a problem he knew he would have to tackle on a large scale on his return. He spent six weeks in Lombardy and Piedmont. In Lombardy, he visited the heads and other works on canals near Milan; he examined the Ticino River from Sesto Calende to its mouth, and along the line of the Naviglio Grande to Buffalora; he visited locks and other works on the Pavia Canal; he was interested in the successful works for the riddance of silt from a canal near Bergamo, and in the bunds in use on the River Po near Mantua and Cremona. But as for learning in Lombardy how to deal with canals in contact with moutain torrents, Cautley was disappointed. The Milanese canals draw their supplies from Lake Maggiore and Lake Como, natural reservoirs which intercept the torrential flow of the rivers Ticino and Adda. By this fortuitous phenomenon, the canals are protected from the effects of the torrents. Cautley was envious, too, of the superior quality of soil in the beds of the canals, unlike the sandy beds in the canals he was building, and the excellent granite from the Bavino quarries, which spared the Italians the necessity of having to make bricks. He saw only one mountain torrent crossing a canal: the River Agogna crossing the Mora Canal, north of Novara. But like all the works he saw in Italy, everything was on a much smaller scale than the works he was contemplating. He was struck by the simplicity of means employed everywhere — simple devices for regulating the supply, raising sluice gates, fixing the gates of locks, and so on — for this concurred with his practice in India of making the works as easy as possible for maintenance by local people. In particular, Cautley felt 'surprise, mixed with a good deal of satisfaction, at the numerous instances in which we, who were entirely separated from all communication with the Italian engineers, had, by the mere process of simple reason, arrived so frequently at precisely the same results, and in so many cases had adopted the same expedients.'[92]

Cautley managed a brief visit to the works at the head of the delta of the Nile before setting sail for Bombay, where he arrived on 14 December 1847. On 11 January 1848, he had resumed charge of the Ganges Canal works.

THE BUILDING OF THE CANAL, 1848-1854

In Cautley's absence, the last negotiations with the government had taken place. A further delay had been occasioned by the government's desire to satisfy itself that irrigation works would not introduce malaria to the Doab via the stagnant pools formed by leakage through the canal banks. Accordingly another committee was appointed on 16 September 1845 to enquire into the causes of unhealthiness in recent years at Kurnal on the Delhi Canal and to report whether the Ganges Canal was likely to injure the health of the people of Doab. The committee was composed of Major Baker, Surgeon T.E. Dempster,[93] and Lieutenant Yule,[94] although in practice Yule took no part in the proceedings. The Report was delayed by the intervention of the Sutlej War in 1845-46, to which Baker and two of his officers were called on service, but it was completed and dated 3 March 1847.[95] It concluded that there had been illness everywhere after the rainy season of 1843, that some of it was exacerbated by poor maintenance of parts of the canal channel, but that there was no reason to think that the Ganges Canal would have a damaging effect on health if certain precautions were taken: for example, that the canal should be built 'within soil,' as far as possible and not carried above ground level in massive embankments as had been done on the Doab Canal.

This Report was presented to Lord Hardinge when he visited the works at Roorkee in March 1847, with the result that on 1 May the Supreme Government ordered the Agra Government to prosecute the work with the utmost diligence. The long period of government reluctance and obstruction was over. Approval was given for irrigation to be the primary object and navigation the secondary; the works between the Ganges and Roorkee were to be as in Cautley's 1845 Report; the precise course of the canal below Rookee was still to be determined; water power was to be made use of wherever possible; and reservoirs and plantations were to be created as on the Jumna Canals.[96]

Baker's plan, explained to the Military Board in a report dated 13 April 1847, was to complete the aqueduct over the Solani river first, to finish masonry works before earthworks since the latter were liable to injury from weather, and to complete the whole project, main and branch lines, simultaneously.[97] The government, now agreeing enthusiastically to the project, authorized a sufficient staff, composed of the Director and seven executive officers, one of whom was solely in charge of organizing materials, the other six being responsible for each of the six divisions. Each executive was to have a staff composed of two assistants, an accountant and two writers (English office), and a native establishment of about fifty. After the rainy season, that is, in November, the officers were to obtain an accurate survey of their districts and gather bricks along the line wherever fuel was available. With the Director's approval, the line would be marked out and the principal works begun, agreeably to Cautley's design. Executives were to send in quarterly accounts to the Director, from which he would make an abstract to send for audit to the Military Board. Quarterly or half-yearly as required, the Director would submit balance-sheets to the government. Instructions, dated 30 September 1847, were given to the executive officers to examine their areas and to prepare estimates regarding the amount of excavation involved.[98]

Despite Baker's absence with two assistants on military service between 1845 and 1846, Cautley returned to find everything now ready to proceed and work under way in the northern division. The sinking of block foundations for the Solani aqueduct had started, work was advancing on the Myapoor dam at the headworks and on some of the bridges and falls. *Chokis* had been built and the first forty miles had either been excavated or were in the process of excavation. Executives had been appointed to five of the six administrative divisions: two executives and four assistants in the northern division; one executive and two assistants in the second and third divisions; an executive and one assistant in the fourth and sixth divisions; while in the fifth division (the Etawah terminal branch) surveying was still being carried out to determine the best line.

For the first one hundred miles, the three projects Cautley had submitted in 1845 were the same, but the progress of the work now revealed the nature of the ground, a sandy substratum beneath clay supersoil, the latter being largely removed by excavation. Doubtless his experience of sand in the bed of the Doab Canal now made him wish to reduce the slope; the Medical Committee's requirement of keeping the canal 'within soil' meant some modifications; and a re-projection of some of the structures and the canal's capacity had become necessary. In 1850 the Agra Government requested a report with revised estimates, which, dated 15 September 1850, shows the scheme which was finally built.[99] The main line ran 180 miles to Nanoon, encompassing the first three administrative divisions. At Nanoon, it split into two branches. The first was 169 miles long and encompassed the fourth and sixth administrative divisions before reaching Cawnpore. The second branch, the Etawah terminal of 165 miles, comprised the fifth division and made its way to the Jumna at Furruckabad. From the main line the Futtigurh branch, 160 miles long and included in the second division, took off at the 50th mile; the Bolundshuhur branch of 90 miles parted at the 110th mile and the Koel branch of 50 miles at the 145th mile, both in the third division. From the point where these last two branches converged, a main irrigation line of 160 miles ran to the Jumna west of the Etawah terminal. Completed, the navigable channel would be over 500 miles long, and the main branches, for irrigation only, about 460 miles (Figure 10).

The work was accomplished in the way that Baker had suggested. Each executive engineer surveyed his own line in detail and built the structures on it according to the Director's design, or to his own design with the Director's approval. Apart from the first 20 miles where the structures were novel and difficult, the main structures were bridges at roughly 3 mile intervals, escapes and cuts to pass off excess water to neighbouring rivers, main water-course heads and falls with attendant works of locks, navigable channel, and mills. Accounts of materials used, wages paid to labourers and progress made were submitted to the Director monthly, as well as returns of printed questionnaires supplying details of engineering construction: examples of these can be seen in the appendices to Cautley's Report of 1860.[100]

The depth of excavation varied according to the terrain. In the northern division, deep cutting was needed to bring the canal through the Peeran Kulleeur ridge to Roorkee, with a mean depth of excavation of 31 feet (Figure 9b). Below Roorkee, the canal passed through several ridges which entailed deep cuttings of, for example, over 67 feet in parts of the stretch between the 40th and 60th miles, and over 44 feet in parts between the 60th and 80th miles. The depth of the channel from the bottom to the top of the bank was 15 feet in the first 50 miles, 12 feet for the next 60 miles, and 13 feet for the next 70 miles. This brought the canal to the bifurcation at Nanoon, and here the comparable depth was 11 feet dropping to 10 feet at the terminals.[101]

The work of excavation was performed by labourers, many of them members of such nomadic tribes as the Oades, who seem to have specialized in this kind of work. The men worked either by contract, usually unwritten, organized in gangs under a leader, or were employed directly, daily or monthly, supervised by the European overseers. Each man on an average moved 83.33 cubic feet per day. Cautley took pleasure in noting that some of the men employed on the Ganges Canal were the children of excavators who had worked for him on the Doab Canal.[102] There were often as many as three or four thousand employed daily under any engineer, digging the earth with large mattocks, putting it in baskets, and leading it away on donkeys. Thomas Login,[103] executive engineer in the northern division, in his account of the work says that on 1 February 1854 he had 7,000 workmen and 500 carts employed on his six miles of canal (Figure 11).[104]

The cost of excavation from Myapoor to Roorkee was between 2 and 4 rupees per 1,000 cubic feet, and below Roorkee 2 rupees or less.[105] In sterling of the day, between 5,000 and 10,000 cubic feet could be moved for £1. Thomas Login, in 1857 giving a Scottish audience some idea of the quantities of earth moved, largely in six years, said it would cover the whole of Edinburgh, 2¼ square miles in area, to a depth of 50 feet.[106]

The earth in the bottom of the channel was spread and pounded and the embankments made by 'puddling,' mixing the earth with water and ramming it to make a water-tight seal. In some stretches Cautley employed light rails on which side-tilt waggons, drawn by men or horses, were used to move the excavated earth; on the Solani works, he briefly introduced a locomotive from England, although its career turned out to be short. The timberwork, metalwork, waggons, rails, and so on, were all made in the workshops set up at Roorkee with the encouragement of James Thomason, whose lively interest in the canal project had done so much to promote it.[107] The need for personnel with training in civil engineering led to the establishment of the Roorkee College of Engineering, opened in 1848, and later renamed Thomason College of Engineering in memory of James Thomason, whose premature death a few months before the opening of the canal was deeply regretted. It has now become the University of Roorkee. Later, Cautley instituted the Cautley Gold Medal to be awarded annually to the best mathematician of the year.[108] His interest in passing on to the Canal Department the benefit of his own and his colleagues' experience is seen in the collection of useful tables which he edited in 1851.[109]

In 1848, after some experimentation with native methods, brickmaking machinery was introduced from England.[110] The structures of the canal were built of brick or *kunkur* where it was available, all the

Figure 11. Sketch of the Ganges Canal works from the bridge at Roorkee, 1849, by Henry Yule, one of the canal officers (from [Yule & Maclagan], *Memoir of Sir William Erskine Baker*, 1882, p. 45)

material being provided by the resources of the canal officers themselves. Trees had to be cut down and the timber carried from the forests for the brick kilns, the bricks had to be made, the lime burned, and the lime and sand mixed for cement or stucco. Login's picture of the number of bricks used on the Solani aqueduct and its approaches was that of a line of bricks which would, if placed end to end, more than encircle the globe.[111]

The engineering of the canal was concerned basically with two problems: to maintain the flow of water in the channel at a velocity which would not damage the banks or cause siltation but would be sufficient to reach all parts of the distribution system; and, as on the Doab Canal, to prevent interference with the system from cross-drainage. The first of these problems had two aspects: the distribution of the quantity available and the slope on which this quantity could most efficiently be carried.

Cautley tackled the problem of distribution by reference to the experience of himself and Baker on the Doab and Delhi Canals. In his Report of 1840 on the Ganges Canal, he stated he had calculated that for these canals a constant discharge of 1 cubic foot per second was sufficient to irrigate 350 beegas of 55 yards square each.[112] It is difficult to know how this discharge figure was calculated, but there are three possibilities. One is that he used a current-meter;[113] another that he made a rough estimate of the velocity by using timed floats thrown into

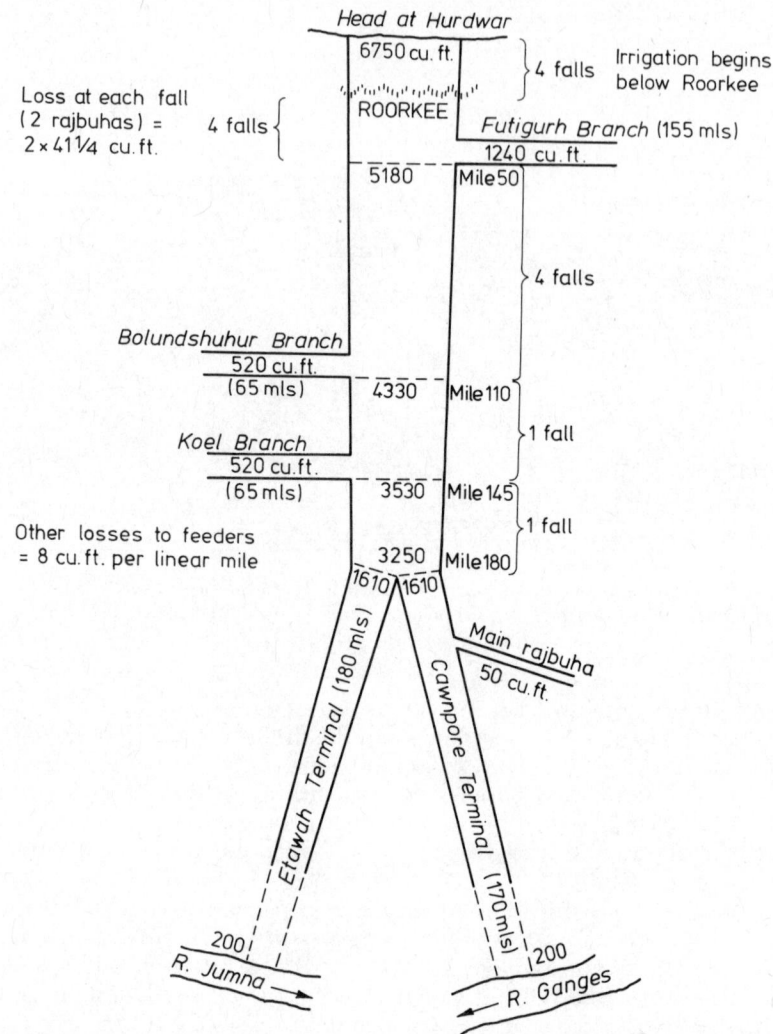

Figure 12. The project of 1850 for the discharge distribution (redrawn from Cautley, *Report on the Ganges Canal Works* ..., 1860, Vol. 1, p. 197)

the stream, a common enough method of gauging flow rates; a third possibility is that he used the formula of the French hydraulician, Dubuat (1734-1809),[114] a formula generally adopted for open-channel flow at this time, and which he was certainly using five years later in the 1845 Report. Records kept over a five-year period of the height of the water on the two water-gauges on the Delhi and Doab Canals, combined with his knowledge of the area irrigated, supplied basic data for the calculations.[115] From it, he was able to conclude that 800 cubic

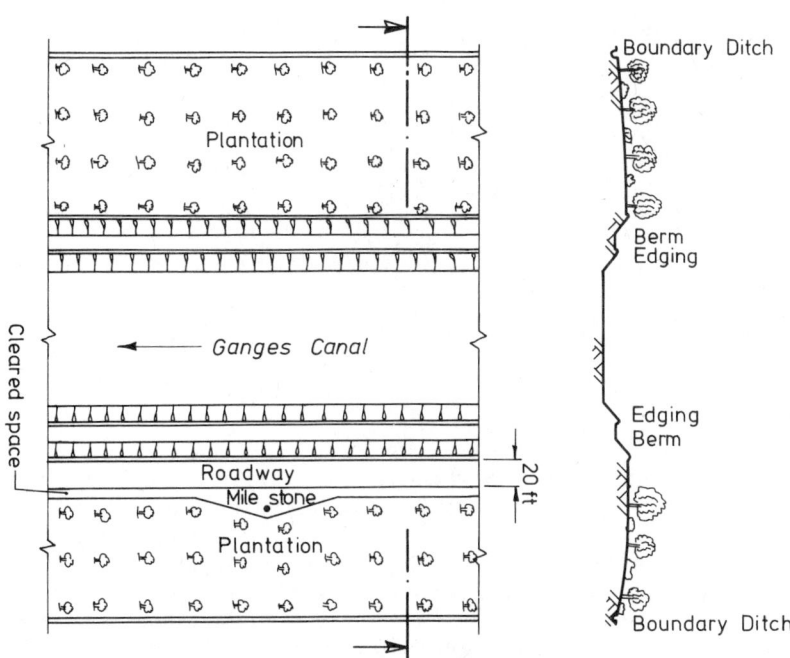

Figure 13. The design of the channel (redrawn from Cautley, *Report on the Ganges Canal Works* ..., 1860, Vol. 3, p. 154)

feet per second was sufficient to irrigate 100 miles in length of canal, assuming that an area from 4 to 5 miles wide on each side would be watered. He therefore judged that of the 8,000 cubic feet per second discharge at Hurdwar, 6,750 cubic feet per second would be sufficient to irrigate over 800 miles of channel length. By varying the trapezoidal section of the channel so that it decreased in width and depth as it approached the terminals, he hoped to deliver water to all parts of the system (Figure 12). The bottom widths would therefore vary from 140 feet at the head to 20 feet at the terminals, and depths correspondingly from 10 feet to 4 feet.[116] Other features of the channel design were a berm, or ledge, at each side raised 12 inches above high water mark, and above the left-hand one a roadway approximately 20 feet wide, built not less than 3 feet above high water level, along which canal staff passed to their duties of maintenance and inspection. The banks were planted with mango trees to provide both shade and produce which could be sold (Figure 13).[117]

Since the slope of the canal bed was the object of much of the later criticism of the canal, it is worth examining Cautley's reasoning about it. There were two basic considerations: the land profile and the maximum allowable velocity. In relation to the first, Cautley's aim was to make the slopes equal on extended lines as far as possible and comparable to the slope of the country. The 1840 Report supplies the data for his conclusion that 18 inches per mile would be suitable for the head of the canal, dropping to 14 inches at the tail, where he hoped that the evil of having the tail on a smaller slope than the rest would not be significant.[118] A summary gives:

	Average slope of country (ft./mile)	Slope of country adjusted (ft./mile)	Slope of canal bed (ft./mile)
Roorkee to Chitowra, 38 miles	2.403	1.851 including 3×7 ft. falls	1.5 with falls
Chitowra to Koel, 119 miles	1.507	1.507	1.5
Koel to Mynpoorie, 78 miles	1.138	1.138	1.166

A slope of 18 inches per mile seemed to him satisfactory since he had found 17.6 inches per mile effective on the Doab Canal in light soil conditions. This view was reinforced in 1845 when he used the formula of Dubuat to calculate the bed velocities that the remodelled slopes of 24 inches per mile on the Doab Canal would give him, which varied between 3.0 and 3.5 feet per second. He could then compare these bed velocities with those calculated for the Ganges Canal on various hypothetical cross-sections of channel. He also checked these bed velocities against calculations made using the formula of Prony, another French hydraulician.[119] They gave him this result:

	V (mean vel.) (ft./sec.)		U (bottom vel.) (ft./sec.)		Slope of canal bed
	Dubuat	Prony	Dubuat	Prony	
at head of canal	4.46	4.15	3.79	3.18	18 ins/mile
at Nanoon bifurcation	3.21	3.05	2.65	2.33	

In 1850, because of the sandy soil exposed by excavation, he decided to reduce the slope by 3 inches per mile, which gave him mean velocities of 4.04 dropping to 3.53 feet per second (presumably using Dubuat's formula), with bottom velocities correspondingly lower.[120]

The later history of the canal, however, demonstrated that these velocities were too high and made it appear that 3.0 feet per second was the maximum safe mean velocity and that 15 inches per mile would have been a satisfactory slope if the depth of water had been less than 5 feet, and not the 6 or 8 feet carried on some parts of the system.[121] James Crofton,[122] asked to report on the canal in 1864, found that the actual mean velocity at the head was 4.7 feet per second and 4.0 feet per second at Nanoon and this with only two-thirds of the supply in the canal. He recommended that the highest mean velocity that should be contemplated for the remodelled canal was 3.0 feet per second in good soil and 2.5 feet per second in poor.[123] It is interesting that in 1870 when the velocities for the Lower Ganges Canal were being considered, a slope of 6 inches per mile was thought 'the greatest declivity which can conveniently be given to the Canal in its upper reaches,' which would create a mean velocity of 2.66 feet per second.[124]

The design of the slope on such a steep fall turned out to be the principal defect of the canal and the source of the later problems that arose. Cautley's reasoning about it was sensible, but his assumption that what had worked well on a small canal could be applied to a large one carrying much greater volumes of water, was mistaken. Moreover, and Cautley evidently failed to appreciate the point, the design formulae so far established for open-channel flow were hardly capable of rendering much assistance. Although they anticipated the correct empirical form, the equations of Dubuat and Prony amongst others did not yet embrace the correct values of the necessary constants and neither had they taken proper account of the influence of surface roughness. Nothing very reliable or comprehensive was in fact to be available until the work of Ganguillet and Kutter (1869) and Robert Manning (1890). In reality, Cautley, working as he was in the 1850s on large channels carrying high flows through sand, was better placed to add data to the search for empirical design formulae than the existing formulae were in a position to guide the design of so singular a project as the Ganges Canal.

The chosen slope was achieved by introducing falls, fourteen in all. Four of them occurred in the first twenty miles. These were basically of the same design as those on the Doab Canal, with the water passing over an ogee curve. The crests of the falls were built at the upper canal bed level and were usually of the full width of the canal. When a large body of water was moving down the canal, the effect was to increase the velocity of the water and to decrease its depth for some distance above the fall. As the water ran down the ogee, its velocity increased even more, wreaking havoc on the flooring and creating a standing wave below the fall with considerable washing of the banks. Some attempt was made in the design to reduce velocity by delivering the water over the ogee into four chambers made by prolonging every alternate bridge-pier on its downstream face. Again, with the wisdom of hindsight, we can now say that some other design would have been preferable, for while ogee falls worked reasonably well on the Doab Canal, they were unsuitable for the vastly increased volumes of water passing over them on the Ganges Canal. When remedial works were undertaken in the 1860s, the crests were raised by a masonry weir and the ogee curve was in some cases converted to a perpendicular drop through gratings into a cistern at the foot of the fall where some of the shock could be absorbed, a type of fall successfully developed on the Bari Doab Canal, built 1850-59.[125]

The second problem, the prevention of interference from cross-drainage, had to be tackled with nothing if not boldness. Water for the canal is diverted from the Ganges at a point 2¼ miles above the town of

Hurdwar, where a branch of the Ganges is divided from the main stream by a wooded island, from whence it travels past Hurdwar to rejoin the parent stream at Kunkul a mile and a half below Hurdwar. By building a temporary spur into the main river, deepening the channel of the branch, and closing off three escape channels in it with embankments and bunds, the water was brought past Hurdwar to a point where the new cut could begin (Figure 14). Here at Myapoor, a regulator was built on the canal and a dam, or weir, on the branch of the river. These were large structures: the regulating-bridge had ten bays of 20 feet each, fitted with shutters, while the dam, built on a large platform, had fifteen openings of 10 feet with their sills 2½ feet above the canal bottom; these sills were removable to give a flush waterway.

The first cross-drainage problem arose 5 miles from the Myapoor regulator where the Ranipoor *rao*, a torrent 100 feet wide on a sandy supersoil with a catchment area 10 miles long and 4½ miles wide, approached the canal by a tortuous route on a slope of 18 feet to the mile. In the original design it had been intended to allow the torrent to fall into the canal by a perpendicular drop of 12 feet in the canal bank, and to escape at the opposite side over a dam, 131 feet wide, provided with gates, and afterwards to flow into an excavated escape channel. Severe flooding in 1849, however, resulted in the disappearance of both structures under sand deposits. Thomas Login,[126] the executive engineer on this section, suggested that the nearest fall should be brought to this point, so that the canal bed could thus be lowered 9 feet, which would allow it to be passed under the *rao*.[127] Cautley agreed to this proposal, and a waterway 200 feet wide was provided for the *rao* on a sub-structure of eight arches, each of 25 foot span, and incorporating a fall of 9 feet. Bunds and embankments were constructed to direct the torrent on to the super-passage.

A few miles farther on, 9½ miles from the regulator, the canal was cut by the Puttri torrent, which, with a catchment area of 70 square miles, approaches the canal in a wide sweep. It was decided in this case also to construct a super-passage and redirect the torrent on to it. Its waterway was 300 feet wide on a sub-structure of nine arches of 25 foot span. The fourth set of falls, which were to have been situated 3 miles below where the Puttri crossed the line were incorporated with the super-passage at the point of crossing. Poor soil containing spring water not far below the surface made construction difficult, and very solid foundations covering an area 331 feet by 285 feet were built.

At about 13 miles from the regulator, a third and larger drainage problem occurred at Dhunowri as the Rutmoo river, with a catchment area of 126 square miles, met the canal on its own level. This river is perennial, though subject to a rise in the wet season. To deal with it, Cautley devised what can be described as a level crossing. The works consisted of cuts and spurs made to direct the river into a new channel, an inlet and dam built across the Rutmoo, a regulating-bridge and bridge on the canal, and drainage works incorporated into the revet-

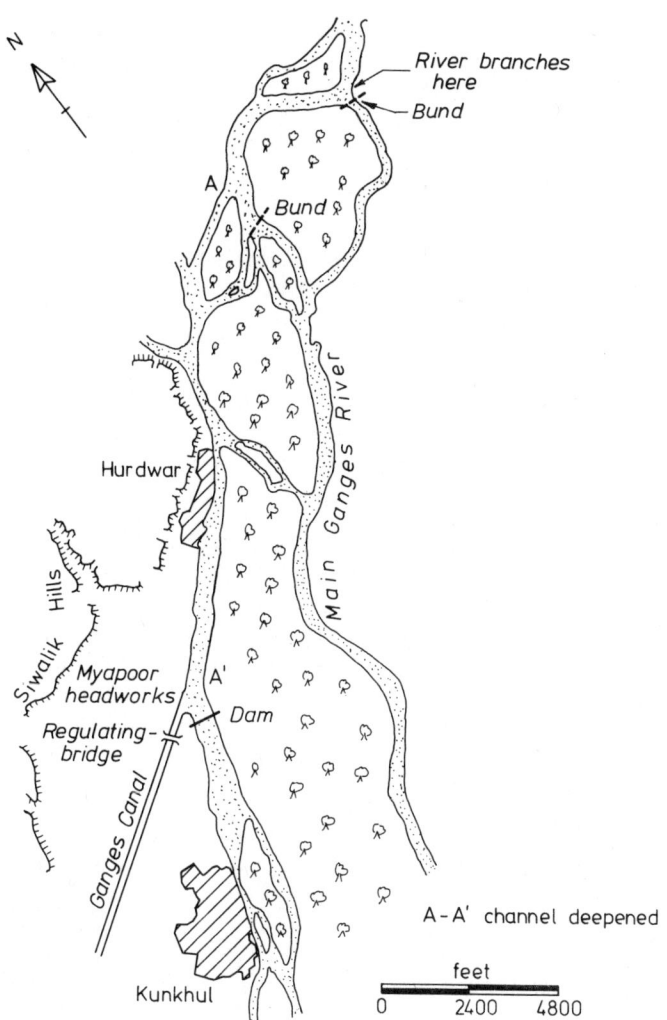

Figure 14. Sketch-map of the headworks of the Ganges Canal, 1854

ments of the curved channel as it approaches the works. The inlet had twenty-nine centre openings 10 feet wide with sills 2 feet above the canal bed; there were fourteen of similar width on each flank and platforms of 17 feet on the extreme flanks, 10 feet above the bed of the canal. The gates in the centre sluices were designed to fall downstream to the pressure of the torrent. Opposite it was a dam composed of forty-seven sluices of 10 feet with sills flush with the canal bed; on each side were five sluices of the same width, but with sills raised to a height of 6 feet; on the extreme flanks, platforms 17 feet long were 10 feet above the canal. This allowed a maximum waterway of 800 feet in flood conditions. Double sluice-gates could be used singly or together to regulate the flood water and the water in the canal. In this they were aided by the regulating-bridge built downstream of the flood regulation works. A double drain of two channels each 7 feet wide with apparatus for opening and shutting passed under the canal and through the foundations of the regulating-bridge and released its water into an escape cut. In this way water could be passed off when the river was not in full flood (Figure 15). All these works were built on massive foundations of cubes of brickwork, some with a base area of 32 × 22 feet and sunk to a minimum of 12 feet.

The troublesome cross-drainage had thus so far been dealt with by passing it over, under and through the canal in structures probably quite novel in India. Cautley may have derived some of the ideas for them from his visit to Italy, but even if that is so, the works in Italy were by no means on the same scale as those on the Ganges Canal.

To cross the last and largest seasonal torrent, the Solani River, with a catchment area of 216 square miles, Cautley had early conceived the idea of an enormous aqueduct (Figure 16). It was, without doubt, the show-piece of the canal, at that time and until the end of the century the longest aqueduct in the world. The Solani valley was almost three miles wide, and Cautley decided the canal should cross it by an aqueduct across the whole distance, with the channel enclosed first in an earthen embankment, then in a masonry aqueduct, and again in an earthen embankment. The two sections of earthen embankment were 10,713 feet and 2,723 feet long, with the masonry aqueduct 1,110 feet long between them, 24 feet above the river bed.

As the canal approached the Solani valley, the works constructed were first of all escapes on the right and left banks, then cattle ghats, then the earthen channel meeting a bridge at Mahewur in ogee curves of brickwork, followed by another section of earthen channel joined to the masonry aqueduct. This was carried on a substructure of fifteen arches, each of 50 foot span. The earthen channel was joined to it at the other side and met the bridge at Roorkee in ogee curves as at Mahewur, followed by earthen channel and cattle ghats. The symmetry of the works was emphasized by the placing of two pairs of brick lions at each end of the works; one pair faced upstream from the Mahewur bridge and one pair downstream from Roorkee, while those on the aqueduct faced each other.

As the designer of the Solani bridge Cautley's preoccupations were to ensure in every way he could against failure of the structure, since the sudden release of such a huge volume of water from the aqueduct would cause enormous damage. Consequently, the masonry aqueduct was built on massively strong foundations, composed of huge cubes of brickwork, many of them 20 feet square, sunk by manual excavation to a depth of 20 feet below the river. Over three hundred of these blocks were used under the aqueduct. As an emergency measure, escapes 100 feet

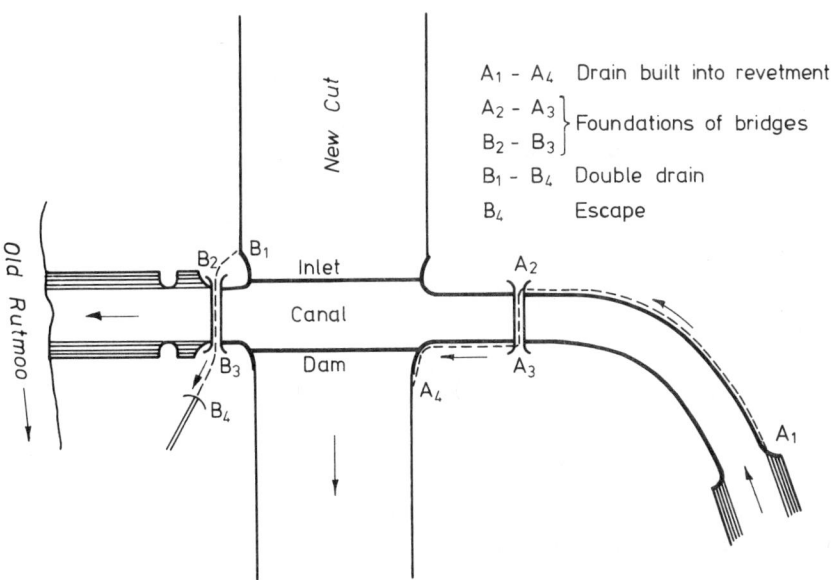

Figure 15. The drainage design at Dhunowri (redrawn from Cautley, *Report on the Ganges Canal Works* ..., 1860, Vol. 2, p. 21)

wide were built in each bank before the start of the works at Mahewur. These were sealed with earth, which could be removed by hand. By closing the regulating-bridge at Dhunowri and using the escapes, the bulk of the canal's flow could be prevented from reaching the aqueduct bridge. Another protective measure was to build the aqueduct channel

Figure 16. View of the Solani Aqueduct (from Cautley, *Report on the Ganges Canal Works* ..., 1860, atlas, frontispiece)

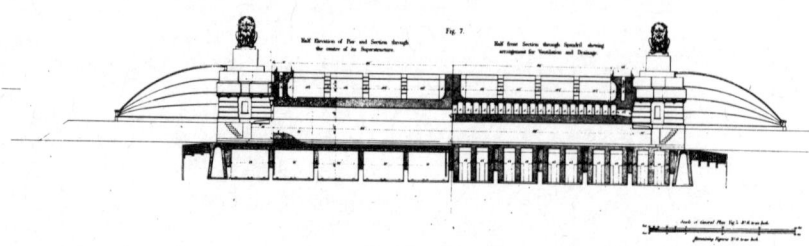

Figure 17. Section through the masonry aqueduct, Solani (from Cautley, *Report on the Ganges Canal Works* ..., 1860, atlas, Plate 27, sheet 2)

in two parts carried on separate piers to allow for uneven settlement (Figure 17). This also permitted one channel to be closed completely for repairs or other reasons, and, Cautley hoped, lessened the effect of wind sweeping across the water whose channel was 172 feet wide on the aqueduct.[128] Great care was taken over the junction of the earthen with the masonry aqueduct, by building below the earthen channel a masonry platform and arch abutting on to the masonry works at the point of contact (Figure 18).[129] To avoid any possibility of leakage, the earthen channel was lined throughout its length by double walls of brickwork, the space filled with earth and an arch thrown across the top of the double wall (Figure 19). The bottom of the channel was filled

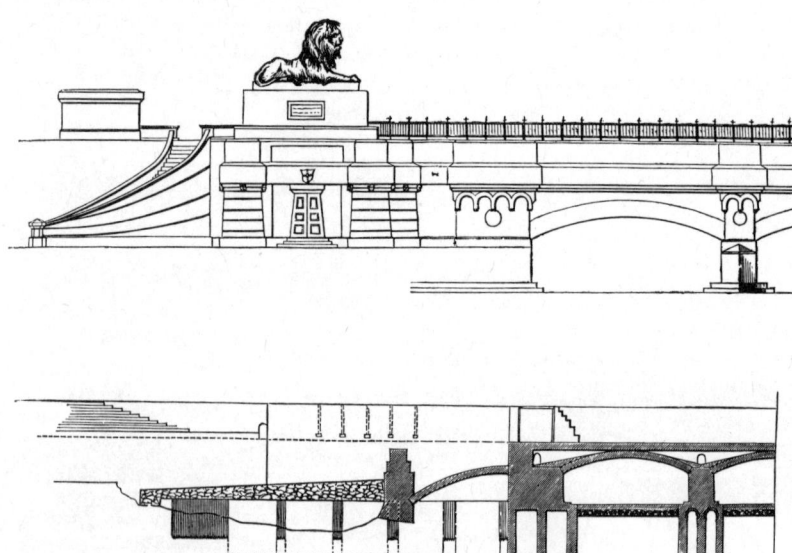

Figure 18. Section showing the junction of the earthen and masonry aqueducts (from *Engineering*, Vol. 1, 1866, p. 377)

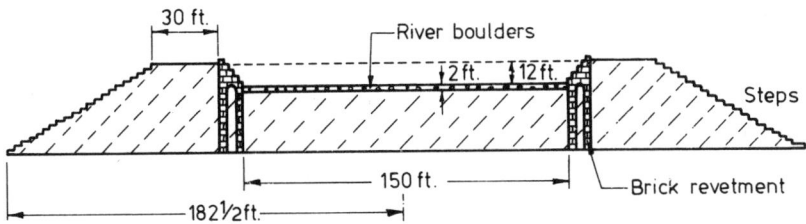

Figure 19. Section through the earthen aqueduct, Solani (redrawn from Cautley, *Report on the Ganges Canal Works* ..., 1860, Vol. 2. p. 434)

with boulders, but these were not to be laid until the action of the current on the bed had been observed for some time.[130]

The construction of the earthen embankment began in 1845-46 on the low ground between the Solani and Mahewur. Trenches 188 feet wide and 5¼ feet deep were dug on either side of the aqueduct line and the earth thrown onto the centre at *a* in Figure 20. The revetments were then built and the spaces *b* and *b'* filled from *a* on which a railway ran; the sides *c* and *c'* were made simultaneously and built to the top of the revetment later. The whole formed a trapezoidal mound with a base 365 feet wide. Light rails on its centre and sides carried side-tilt waggons full of additional earth from the excavation of the Peeran Kulleeur ridge. Work began at the Roorkee end simultaneously. Where the earthen channel met the masonry aqueduct, the height of the embankment was 37¼ feet on the Mahewur side and 36¾ feet on the Roorkee side. The inside of the channel was revetted with steps and the exterior slopes finished with steps 4 feet wide every 1,000 feet. The bottom width inside the channel was 150 feet expanding with ogee curves to 172 feet at its junction with the masonry channel.

Enormous quantities of materials and labour were used in the construction of the Solani aqueduct. Cautley gives the final cost as 3,286,812 rupees (£328,680),[131] which compares with his 1840 estimate of 1,000,113 rupees (£100,000). The total cost of the canal was of the order of £1½ million,[132] one third of which was spent on the northern division, which is not far removed from his 1850 estimate for that section of 5,747,846 rupees (£574,785).

Below Roorkee there were fewer problems. The land of the second and third divisions was on a fairly steep slope and intersected by low sand-hills. This meant that much of the canal had to be contained in

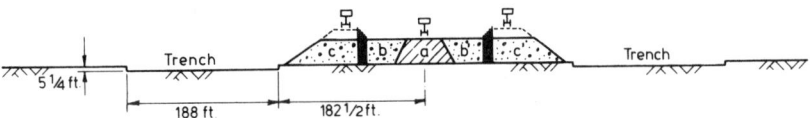

Figure 20. Method of building the earthen aqueduct, Solani (redrawn from Cautley, *Report on the Ganges Canal Works* ..., Vol. 2, p. 436)

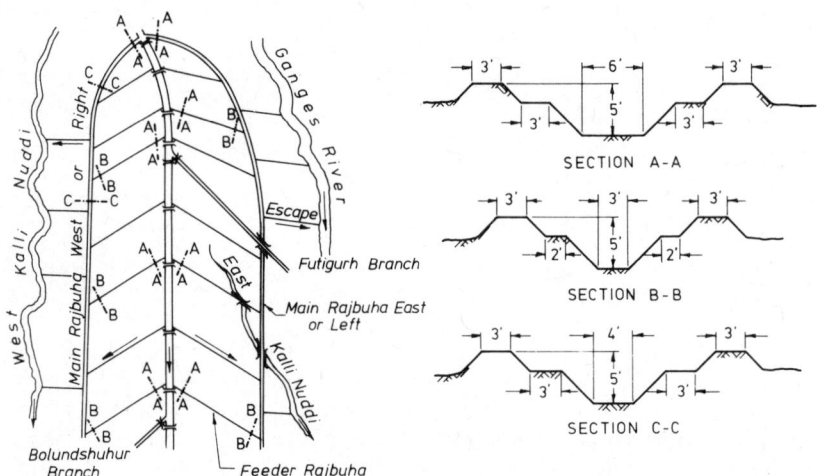

Figure 21. Diagram showing the method of irrigation distribution, Ganges Canal, 1854 (redrawn from Cautley, *Report on the Ganges Canal Works* ..., Vol. 1, p. 197)

sandy soil, and the slope overcome by ten falls, eight of 8 feet and two of 5 feet. In the two terminal lines, the fall of the land was almost too slight, and there were problems associated with deflecting the drainage of the various rivers of the plain away from the canal. Distribution was organized as on the Doab Canal, with water taken off on this canal at 3 mile intervals (Figure 21).

A bad working season of cold weather rains in 1850-51 was followed by good progress in the next two years, and the canal was far enough advanced to pass water to the southern districts. The main trunk and Cawnpore Terminal were finished, the first 60 miles of the Etawah Terminal were ready, and work had begun on the three branch lines.

The official opening on Saturday, 8 April 1854, was given all the splendour of a royal ceremonial occasion. It was at first intended to admit water formally at the Dhunowri regulating-bridge with the official party accompanying it on its way to the Solani aqueduct in special railway coaches along the banks.[133] This plan was thought impracticable in the end, and the official opening was planned with the admission of water on to the Solani aqueduct. Accordingly, on 1 April, two of the canal officers breached the bund which separated the Ganges from the cutting connecting the river bed with the canal and admitted a large body of water nearly 3 feet deep. The water travelled slowly, and reached the aqueduct on 7 April, where efforts were made to prevent it overtopping the gates before the official opening, by releasing some of it through the mill sluices.

The ceremony was performed at six o'clock in the morning by John Russell Colvin, Lieutenant-Governor of the North-West Provinces.[134]

He undid the levers which kept the canal gates closed, which was the signal for all eight gates to be thrown open to the accompaniment of the national anthem and the firing of salvos.

Cautley left India as soon as the canal was officially opened. His achievement was honoured by the City of Calcutta in placing a bust of him in the Town Hall,[135] and by the Governor-General's ordering a salute of thirteen guns to be fired from the ramparts of Fort William as his boat made its way down the river to join the steamer that would take him to England.[136]

THE QUARREL WITH SIR ARTHUR COTTON, 1863-1865

Cautley may well have thought in his retirement in England that his preoccupation with the Ganges Canal, except in the most general way, was over. In the first years after his return, he was busy completing his *Report on the Ganges Canal Works*, in three volumes with folio atlas of plans, published in 1860. Messrs. Smith and Elder had offered to publish five hundred copies at a total cost of £6,350,[137] which no doubt accounts for its scarcity today.

Late in 1863, however, Cautley was greatly irritated by a public argument which arose about the design of the Ganges Canal. In the first few years after the canal's opening problems began to arise, springing mainly from the chosen slope of 15 inches per mile, which, as had already been suggested, was much too high. The velocity of the water caused scour and damage to some of the masonry structures, particularly the falls. It became clear that a certain amount of remodelling would have to be done. In addition, the canal did not in its first years show much financial profit, partly because of the damage done to it in the Indian Mutiny in 1857 and partly because its branches and distributary system were not yet complete. Following a bad famine in 1860-61, Colonel Baird Smith's[138] *Famine Report* showed that the canal had proved inadequate to prevent starvation, largely because the temporary nature of the headworks had made it difficult to maintain the canal supply when the levels in the river dropped.[139] Smith, as former Director of the Ganges Canal in succession to Cautley, was in fact using this occasion to urge the Government to spend money on permanent headworks. In the hands of Cautley's adversaries this statement could be made to appear a criticism of the design of the canal as well as its performance.

Cautley's chief protagonist was Sir Arthur Cotton,[140] former engineer of the great deltaic works on the rivers Cauvery, Godavari and Krishna in Madras. In 1863, Cotton was sent by the East India Irrigation Company to plan an irrigation scheme at Behar. While he was there he visited the Ganges Canal, and reported on the expenditure that would be required to put it into more efficient running order so that the Company could offer to purchase it from the government. The resulting *Private Memorandum upon the Ganges Canal*,[141] as well as his address, later printed, to the Calcutta Chamber of Commerce on 7

May 1863,[142] soon became public. Cautley, feeling that justice as well as his duty to the government required that he should answer the charges made against him, published and had printed for private circulation a pamphlet, which became the first of several produced by him and Cotton.[143]

Cotton had accused him in the *Private Memorandum* of making nineteen mistakes in the projection of the canal, five of them fundamental; the first four, it was claimed, resulted in the canal having cost three times what it need have done.

The five major accusations were that the head of the canal was too far up the river, on a tract of great declivity with heavy drainage problems; that the canal was cut below the surface of the country, entailing unnecessary excavation; that the works were built of brick while suitable stone was procurable in the sub-Himalayas; that the whole of the water was admitted at the head and conveyed three hundred and fifty miles, while some of it might have been obtained at a sufficient level only fifty or a hundred miles from the land it would irrigate; and that there was no permanent head. The fourteen minor mistakes were mainly concerned with the canal's effectiveness for navigation — details concerning the falls, the bridges, the lock channels, and so on, and the arbitrary assertion that the Solani masonry aqueduct need only have been one-third the breadth of the canal in the earthen aqueduct and half the breadth of the river below. He also observed that the fall of between 15 and 12 inches per mile and depth of water of 10 feet gave a current between 2½ and 2¾ miles per hour, too great for the bed, the banks, and effective navigation.[144]

In his *Reply to Statements*, Cautley dealt patiently with each of these points, admitting freely that he had been wrong about the slope, but defending himself on the other points.[145] Cotton's insistence that it would have been better to take the head of the canal from a point near Sookurtal, 95 miles below Hurdwar at the confluence of the Solani and the Ganges, by building a weir across the Ganges, was the main point of contention. Cautley observed:

> His [Cotton's] experience, great as it is, is connected with rivers of an entirely different description to that of the Ganges in its debouche from the Sewaliks. Here we have heavy slopes with large masses of water pouring down with overwhelming violence; there he had much larger bodies of water, but on very much smaller slopes in connection with a true delta.[146]

He defended the depth of excavation by arguing that the porous nature of the soil made extensive use of embankments dangerous. Cotton's claim that he had seen good stone in Hurdwar did not alter the fact that the Sewalik Hills contained stone of very unequal quality, and it would have been very expensive to quarry it and sort it over for use on the canal works. To the charge that all the water was admitted at one point, Cautley replied, 'It will be found more easy to propose weirs and dams on the sandy tracts of the Ganges and Jumna than to execute them.'[147] He agreed that a permanent headworks was desirable, but explained that it had not been executed in the original project for reasons of economy, although he had always envisaged that it might be necessary.[148]

The tone of this pamphlet was courteous, and Cautley hoped that the matter might rest there, but almost immediately Cotton produced his *Observations ... on the foregoing Reply*. A joint publication by both engineers contained Cotton's original memorandum, Cautley's reply to it, Cotton's short memorandum, and his *Observations*.[149] In the *Observations* he claimed that he did not know how his original *Report* came to be printed, but that he was, in any event, only acting for the public good. He insisted on his claim that the works in Madras were built on sand, that there was no difference between the Madras rivers and the Ganges except that the Madras rivers were larger, accused Cautley of lacking experience, and reiterated some of his earlier assertions.[150]

This flat denial of his defence led Cautley to a further statement, in which he wrote a detailed essay on the nature of the rivers in North-West India, where he accounted for the headworks of the Jumna and Ganges canals being placed where they were.[151] He felt he had to defend his fellow engineers against Cotton's 'idle calumny.'[152] His main defence lay in showing that the rivers of Northern India were very different in character from those of the deltas; the Northern engineers were taking the river near its head and making use of the natural fall of the country, while the Madras engineers took the foot of a rapid, elevated it artificially by a dam, and then forced it into channels. He described an unsuccessful attempt made in 1828 on the Western Jumna Canal to establish a new head lower down near Kurnal, but it had been swept away in floods, and had convinced him that the northern engineers were right in taking the head high up.[153] His calculations on Cotton's proposed dam across the Ganges at Sookurtal led him to suppose the dam would have to be between 1 and 1½ miles long and at least 15 feet above the dry weather surface of the river, and that it would be very costly.[154] To Cotton's statement,

> As to the *practicability* of building a weir, of course I did not require any detailed information on that; I saw the Ganges in many places, and found that it was *just of the same character as our rivers*, and I know of course *that what had been done in Madras in many places could be done there*,[155]

Cautley replied tartly,

> The fact is, if everything is to be taken 'of course,' and if neither the breadth of the k'hadir, nor the slope of the river, nor any of the hydrographical details, are worthy of consideration, an engineer may jump to any conclusion.[156]

Cotton's *Private Memorandum*, he said, had been used by a public company with his cognizance as a damaging public document, 'in order to depreciate, in the eyes of the Government, the property for which they [the East India Irrigation Company] were bidding.'[157]

The matter did not rest there, for Cotton clearly enjoyed the battle and produced a *Reply* basically insisting on the same points and taunting Cautley with ignorance and inexperience:

> The utmost that Sir Proby Cautley can say is, 'I don't think'; but from thirty years actual experience it is no presumption in me to say 'I know.'[158]

This controversy had now raged for a year, and Cautley, seeing that it was useless to go on arguing with Cotton, wrote *A Valedictory Note*, which he hoped would end the dicussion.[159] In this *Note* he concentrated on two points, both connected with Cotton's manipulation of figures for his own arguments. But as his pamphlet was going to press, Cautley was roused by a biased article in *The Times*, which summarized the argument to date and came out in favour of Cotton.[160] Cautley was extremely angry and upset, and in a separate statement at the end of the book did his best to defend himself again, and to correct assertions which were actually untrue and had been twisted to present him in the most unfavourable light.

The affair rumbled on during 1865. Cotton produced a *Reply* to Cautley's *Valedictory Note*,[161] and these were reviewed in *The Times*.[162] Meanwhile, the Public Works Department in India had already in February 1864 invited Captain James Crofton, Officiating Superintendent of the Eastern Jumna Canal, to submit a report on the Ganges Canal with estimates and plans for remodelling it.[163] This he did, in November 1864, suggesting various measures for reducing the slope and protecting the masonry structures.[164] To Cautley's relief, it passed through the Council of India meetings in the following June. Cotton was furious that his advice had been disregarded, and in two letters to *The Times*, attacked Crofton's report.[165]

A Dispatch of 1 March 1865 of the Governor-General of India in Council to the Secretary of State for India exonerated Cautley from any blame. It stated,

> Whatever be the present ascertained defects of the Ganges Canal, the claims of Sir Proby Cautley to the consideration of the Government of India for his eminent services are, in our estimation, in no way diminished, and his title to honour as an Engineer still remains of the highest order.[166]

To satisfy public opinion, however, the Governor-General by his order of 24 February 1866 set up a committee under Colonel Commandant Edward Lawford to decide between Crofton's scheme or Cotton's project.[167] The findings of the Committee were reviewed in *The Times* in September.[168] The Committee found in favour of Crofton's plan supporting Cautley's assertion that the Madras rivers and the Ganges could not be treated in the same way.

Some of Cautley's intense irritation is seen in a handful of letters he wrote to Crofton between 1864 and 1866, which have survived.[169] On 26 October 1865 he wrote sadly, 'Whenever therefore the Ganges Canal is mentioned by The Times it will always be in abuse and vituperation.'[170] He kept Crofton informed of Cotton's activities and of the articles in *The Times*, which he believed were written by Mr. Dickinson, Secretary of the Indian Reform Association, backed up by Sir Arthur Cotton, and when Crofton's estimates for remodelling were submitted wrote to him gratefully,

> Nothing could have been better; you have treated me with more leniency than I should in all probability have treated myself.... The defect was this infernal slope, and the way in which this has been referred to both by you, and the Government in its report, deserves my warmest gratitude, when looking at the malignant way in which Sir Arthur Cotton has behaved in his criticism of my project.[171]

When the Report of the Ganges Canal Committee came out, Cotton was prompted to publish another abusive pamphlet, in the form of a long letter addressed to the Under-Secretary of State for India.[172] The affair finally subsided into silence. Later events have shown that Cotton was right in principle that dams could be placed across North-Western rivers lower down in their course, as was done for the headworks of the Lower Ganges Canal at Narora and for the Agra Canal at Okhla, but he had seriously underestimated the cost of building such a dam, and Cautley was right in the economic considerations which led him to make the head at Hurdwar, apart from his intention of bringing water to the area in the northern tracts.

This public quarrel cast a shadow over the careers of both the protagonists, but it demonstrates to us now the empirical nature of their work, the controversial character of much of the hydraulic theory discussed and how little of it was as yet established on firm principles. That Arthur Cotton should have attacked the canal officers of the North-West Provinces so viciously is the sadder when one reads the generous description of Cotton by a northern canal officer writing in 1853 of the Madras deltaic works, 'A natural genius for Civil Engineering, large acquired knowledge, singular professional daring, a strong will, and a perseverance that no obstacles could withstand, were combined in Colonel Cotton's character, and marked him out among his associates as the man best qualified for carrying into effect the novel plans that were now [in 1834] being entertained.'[173]

By the time he left India, Cautley had been there, apart from one furlough, for almost thirty-five years and had devoted nearly thirty of them to hydraulic works. The canals he designed were not without faults, but these faults were not irremediable and the basic design of the canals is acknowledged to be sound and to have stood the test of time. Adjustments were made to the levels and sectional area of the Doab Canal in the 1860s, and permanent headworks were built at Tajewala in 1875.[174] Remedial work was carried out on the Ganges Canal: protective works were added to the fourteen falls, their crests were raised to reduce the slope of the canal and additional falls built with the same object; in 1920 permanent works replaced the temporary ones at the point where the river is diverted into the branch; a 'surplussing dam' was built on the channel opposite Hurdwar in 1897 to divert surplus water from the supply channel back into the river; the dam at Myapoor was enlarged and improved; and parts of the distribution were realigned.[175] Between 1930 and 1956 eight power-stations were constructed making use of the falls.[176] The canal was extended in 1878 when the Lower Ganges Canal was opened, thereby adding over 500 miles of channel and increasing the mileage of the distributory system from 2,700 to over 5,000 miles. The two systems together irrigate more than 2½ million acres.[177]

There is no doubt that the Ganges Canal blazed the trail for later work on perennial canals in India. In the next few decades many works were undertaken: in the 1850s the Bari Doab Canal in the Punjab and the Krishna Delta system in Madras; in the 1860s, four canal systems in Bombay and two in Bengal; in the 1870s the Sone Canals in Bengal, the Lower Ganges and Agra Canals in the North-Western Provinces, and the Mutha and Nira systems in Bombay. By the end of the century the Sirhind and the Chenab Canals in the Punjab were added. This century has added some of the largest irrigation schemes in the world. Nor is the work finished yet.

Moreover, the concept of large scale irrigation spread from India to other countries, in some instances conveyed there by men who had learned their skills in the Public Works Department of India. An outstanding example is Sir William Willcocks, by some thought the greatest irrigation engineer of all.[178] Trained at the Roorkee College of Engineering, he spent his early career in the Public Works Department before the course of his life took him first to Egypt, where he designed and built the Aswan Dam in 1898, and then to Mesopotamia, where he constructed in 1911 one of the greatest irrigation systems the world has ever known which waters an area of 3½ million acres.

Cautley combined immense intellectual power with great administrative ability and the courage to try out novel ideas. His reports on the Eastern Jumna and on the Ganges Canal works formed for many years the basis for the teaching on irrigation at the Thomason College of Engineering at Roorkee.[179] They speak more clearly than anything else of the sheer volume of work undertaken and the attention given to detail, while at the same time showing an intrinsic modesty. Without the piety of many Victorians, Cautley was imbued with the same notions of duty to country but with a childlike satisfaction in what he had achieved. In a letter in 1839 to his friend Falconer, he wrote,

I am very busy — making up for time lost during the cold weather — and am illuminating the Zemendars [landowners] in this neighbourhood on the art of Rajbuha manufacturing — much to their satisfaction — & a good deal to my own too — for after all the pleasure of giving a running stream of water on lands, where water has not been seen before — & the results in the sheets of splendid crops that we of the spade and shovel scatter on the face of this Country — is almost as great as that of collecting Mastodons or Hippotamus remains.[180]

There are no memorials to Cautley in this country and few in India,[181] but, as an Indian writer observes, 'what memorial could be as magnificent as the canal itself?'[182] And Richard Baird Smith, writing in 1855, voices the same sentiment but with a wider application,

Our canals of irrigation will probably be the most permanent records of our dominion in the East; for if the lapse of five centuries was insufficient to obliterate the canals of the Mahommedan dynasty, ours are certainly capable of a far more protracted existence.[183]

Notes

I am grateful to colleagues at Imperial College for useful discussions; in particular to Professor A.W. Skempton, Dr. N.A.F. Smith and Mr. Paul Minton. I also acknowledge the help given me by staff in the India Office Library and the library of the Institution of Civil Engineers. Tracings of the figures were made by Ursula Schüler. Figures 1, 2a and 2b are reproduced from F. Newhouse and others, *Irrigation in Egypt and the Sudan, the Tigris and Euphrates Basin, India and Pakistan*, London 1950, by courtesy of the British Council.

1. A full account of Cautley's life is given in Joyce Brown, 'A memoir of Sir Proby Cautley, F.R.S. (1802-1871), engineer and palaeontologist' (in preparation).
2. The climate and physical features of India are described in C.C. Scott-Moncrieff and others, *Report of the Indian Irrigation Commission, 1901-1903*, London 1903; in Frederic Newhouse, M.G. Ionides and Gerald Lacey, *Irrigation in Egypt and the Sudan, the Tigris and Euphrates Basin, India and Pakistan*, London 1950, pp. 33-8; and in R.B. Buckley, *The Irrigation Works of India*, 2nd ed. London 1905, pp. 1-2, 5-6.
3. The spelling of Indian place names is as given by nineteenth century writers. A list giving modern spellings of principal place names mentioned appears at the end of the Notes.
4. See Buckley, *Irrigation Works*, Chapters 1, 2 and 5, Note 2.
5. Scott-Moncrieff and others, *Indian Irrigation Commission*, Part 1, p. 12, Note 2.
6. Ibid., Part 1, p. 11.
7. Corelli Barnett, *Britain and Her Army, 1509-1970*, London 1970, pp. 275-6; Lord Birdwood, 'The story of the Indian army', *Journal of the Royal Society of Arts*, Vol. 101, 1952-53, pp. 44-55.
8. See William Fidler, *The East India Company: a Memorandum . . .* London 1873; F.W. Stubbs, *History of the Bengal Artillery*, London 1877.
9. Sir Henry Hardinge (1785-1856), first Viscount Hardinge of Lahore, Governor-General of India, 1844-47.
10. James Andrew Broun Ramsay, tenth Earl and first Marquis of Dalhousie (1812-1860), Governor-General of India, 1848-56.

11. See Central Board of Irrigation and Power, *Development of Irrigation in India*, New Delhi 1965, pp. 44-8, 58-75; R.B. Buckley, *The Irrigation Works of India, and their Financial Results*, London 1880.

12. Robert Smith (1787-1873), Bengal Engineers, 1805-30. Colonel; C.B. 1822-30, Garrison Engineer and Executive Officer, Delhi, and 1824-30, Superintendent of the Doab Canal.

13. A history of Addiscombe 1809-60, complied in the India Office Military Department . . ., India Officer Library, L/MIL/9/357, f. 18.

14. H.M. Vibart, *Addiscombe. Its Heroes and Men of Note*, London 1894, p. 154.

15. William Mudge (1762-1820), Major-General, Royal Artillery; F.R.S. 1798, Director of Trigonometrical Survey; 1809, appointed Lt. Governor of the Royal Military Academy, Woolwich; 1810-20, Public Examiner at Addiscombe.

16. Sir Charles William Pasley (1780-1861), General, Royal Engineers; F.R.S. 1812-41, Director of Royal Engineer Establishment at Chatham for field instruction; 1841-46, Inspector-general of Railways; 1839-55, Public Examiner at Addiscombe.

17. Seminary Committee Reports, 1815-20. India Office Library, L/MIL/1/11, Report No. 85.

18. V.C.P. Hodson, ed., *List of the Officers of the Bengal Army, 1758-1834*, London 1927-28, Part 1, p. 322.

19. History of Addiscombe, L/MIL/9/357, f. 24, Note 13.

20. Mildred Archer, 'An artist engineer — Colonel Robert Smith in India (1805-1830),' *The Connoisseur*, February 1972, pp. 79-88; Archer, *British Drawings in the India Office Library*, London 1969, i, pp. 317-23; Archer, *Company Drawings in the India Office Library*, London 1972, p. 194.

21. John Colvin (1794-1871), Bengal Engineers, 1810-39. Colonel; C.B. 1827-36, Superintendent of Canals, Delhi Territory. Major [John] Colvin, 'On the restoration of the ancient canals in the Delhi territory,' *Journal of the Asiatic Society of Bengal*, Vol. 2, 1833, p. 116.

22. Ibid., p. 118.

23. George Rodney Blane (1791-1821), Bengal Engineers, 1806-21. Captain. 1817-21, Superintendent of Canals in the Delhi Territory.

24. William Erskine Baker (1808-1881), Bengal Engineers, 1826-77. General; K.C.B. 1829-36, assistant on the Delhi Canal; 1836-45, Superintendent of Delhi Canals, and of Sind canals; 1845-48, Director of Ganges Canal; 1854-55, Secretary to the Government of India in the Public Works Dept.; 1859-61, Military Secretary at the India Office; 1861-75 Member of the Council of India.

25. William Baker, *Memoranda on the Western Jumna Canals, in the North-Western Provinces of the Bengal Presidency*, London 1849.

26. Henry Debudé (or de Budé) (1800-1843). Bengal Engineers, 1818-1843. Major. 1823-26, assistant of the Delhi Canal; 1831-35, Garrison and Executive Engineer, Delhi; 1835-36, Superintending Engineer in Public Works Department; 1841-43, Secretary to Military Board.

27. P.T. Cautley, *Notes and Memoranda on the Eastern Jumna, or Doab Canal, and on the Water Courses in the Deyra Doon*, Roorkee 1853, pp. 5-7. [Hereafter cited *Notes and Mem.*]

28. [P.T. Cautley], Doab Canal Sketches, c. 1833. Library of the Department of Civil Engineering, Imperial College, London S.W.7.

29. *Notes and Mem.*, pp. 18, 25-6.

30. P.T. Cautley, 'A description of the use of wells for foundations,' *Journal of the Asiatic Society of Bengal*, Vol. 8, 1839, pp. 47-64.

31. *Notes and Mem.*, pp. 41-2.

32. Ibid., p. 172.

33. Ibid., p. 10.

34. P.T. Cautley, *Report on the Ganges Canal Works: from their commencement until the opening of the Canal in 1854*, London 1860, 3 vols and folio atlas [hereafter cited G.C.W.], Vol. 1, pp. 16-17, Vol. 2, pp. 617-19.

35. Richard Baird Smith (1818-1861). Madras Engineers, 1836-38; Bengal Engineers, 1839-61. Colonel. 1840-43, assistant on Doab Canal; 1843-50, Superintendent of Doab Canal; 1854-59, Superintendent of Canals of the North-Western Provinces and Director of

Ganges Canal Works; 1859-61, Secretary to Government of India in Public Works Department.
36. *Notes and Mem.*, pp. 9-10; Appendix C, pp. 200-201.
37. Ibid., pp. 37-8, 40-2.
38. T. Login, 'On the benefits of irrigation in India, and on the proper construction of irrigating canals', *Minutes of Proceedings of the Institution of Civil Engineers*, Vol. 27, 1868, p. 475.
39. Printed as Appendix A in *Notes and Mem.*, pp. 167-84.
40. *Notes and Mem.*, pp. 176-9.
41. W.E. Morton, *Notes on the Levels of the Eastern Jumna Canal...*, Agra 1852, p. 9.
42. P.T. Cautley, 'Discovery of an ancient town near Behut, in the Doab', *Journal of the Asiatic Society of Bengal*, Vol. 3, 1834, pp. 43-4, 221-7.
43. Buckley, *Irrigation Works*, pp. 136-7, Note 2.
44. Anon, 'Description of the regulating dam-sluices of the Doab Canal', *Journal of the Asiatic Society of Bengal*, Vol. 2, 1833, pp. 454-6.
45. *Notes and Mem.*, pp. 34-5, 48-54.
46. Ibid., pp. 131-6.
47. R. Baird Smith, *Italian Irrigation, being a Report on the Agricultural Canals of Piedmont and Lombardy*, 2nd ed., Edinburgh and London 1855, p. 339.
48. *Notes and Mem.*, Appendix D., pp. 202-11.
49. Baird Smith, *Italian Irrigation*, p. 345, Note 47.
50. [Cautley], Doab Canal Sketches, f. 5, Note 28.
51. Robert Cornelis Napier (1810-1890). Bengal Engineers, 1826-90, Field-Marshall, 1st Baron Napier of Magdala, G.C.B.; G.C.S. 1831-36, assistant on the Doab Canal; 1849-56, Chief Engineer, Punjab; 1870-76, Commander-in-Chief in India; 1876-83, Governor of Gibraltar; 1887-90, Constable of the Tower of London.
52. *Notes and Mem.*, p. 141.
53. Ibid., Appendix C. pp. 200-201.
54. Ibid., p. 151.
55. Ibid., p. 150.
56. Ibid., p. 151.
57. Morton, *Eastern Jumna Canal*, Note 41.
58. Baird Smith, *Italian Irrigation*, pp. 340, 350, Note 47; and Baird Smith, *The Special Commissioner's Final Report on the Famine of 1860-61*, Calcutta 1861, pp. 31-2.
59. Revenue returns are given annually in *Statement exhibiting the Moral and Material Progress and Condition of India, 1859-1935*.
60. P.T. Cautley, *Notes and Memoranda on the Water Courses in the Deyra Doon, North Western Provinces*, Calcutta 1845. This is also printed in *Notes and Mem.*, pp. 221-75.
61. Baird Smith, *Italian Irrigation*, p. 351, Note 47.
62. Henry Kirke (1806-1857). 12th Native Infantry, 1822-57. Brevet-Major. 1841-48, Superintendent of Watercourses in the Dun.
63. Mentioned by R.H. Phillimore, ed., *Historical Records of the Survey of India*, Vol. 3, Dehra Dun, 1954, p. 269.
64. P.T. Cautley, 'On the proposed formation of a canal of irrigation from the Jumna, in the Dhera Dun,' *Journal of the Asiatic Society of Bengal*, Vol. 11, pt. 2, 1842, pp. 761-75.
65. See above, Note 34.
66. P.T. Cautley, *Report on the Central Doab Canal*, Allahabad [1841].
67. P.T. Cautley, *Report on the Ganges Canal, from Hurdwar to Cawnpore and Allahabad*, Calcutta 1845, and P.T. Cautley, *Plans and maps to illustrate Report on the Ganges Canal from Hurdwar to Cawnpore and Allahabad*, London 1846.
68. [P.T. Cautley], *Estimate of the Probable Expense to be incurred in constructing the Ganges Canal Works, including the main trunk line, the Cawnpoor and Etawah Forks, and the Futehgurh, Bolundshuhur and Coel Branches*, Umballa 1850.
69. G.C.W., Vol. 1, pp. 15-16.
70. George Eden, Earl of Auckland (1784-1849). Governor-General of India, 1835-41.
71. See above, Note 66.
72. See above, Note 66, Second section, pp. 1-23.

73. Frederick Abbott (1805-1892). Bengal Engineers, 1823-47. Major-General; Knight Bachelor; C.B. 1841-45, Superintending Engineer, North Western Provinces; 1850-61, Lieutenant-Governor of Addiscombe.

74. See above, Note. 24.

75. [F. Abbott], *Report of the Special Committee, appointed to examine the project of the Ganges Canal*, Agra 1842.

76. Thomas Campbell Robertson (1789-1863). 1840-43, Lieutenant-Governor of the North Western Provinces.

77. G.C.W., Vol. 1, p. 31.

78. G.C.W., Vol. 1, p. 33.

79. Richard Strachey (1817-1903). Bombay Engineers, 1836; Bengal Engineers, 1836-75. Lt. General; C.S.I.; G.C.S.I.; F.R.S. 1843-46, Executive Engineer, Ganges Canal; 1862-65, Secretary to Government of India, Public Works Dept.; 1866-69, Inspector General of Irrigation; 1869-71, Governor General's Legislative Council; 1875-89, Council of India.

80. See above, Note 34.

81. James Thomason (1804-1853). 1843-53, Lieutenant-Governor of the North Western Provinces.

82. *Calcutta Review*, Vol. 21, 1853, p. 499.

83. Ibid., pp. 499-500.

84. See above, Note 9.

85. See above, Note 67.

86. See above, Note 67, p. vi.

87. See above, Note 67, p. xiii.

88. Henry Marion Durand (1812-1871). Bengal Engineers, 1829-71. Major-General; K.C.S.I.; C.B. 1832-36, assistant on the Delhi Canal; 1842-44, private secretary to Governor-General; 1861, Foreign Secretary in India; 1870-71, Lieutenant-Governor of the Punjab.

89. See above, Note 1.

90. G.C.W., Vol. 1, pp. 99-104.

91. Unidentified newspaper obituary in scrap-book belonging to the Royal Society, London.

92. G.C.W., Vol. 1, p. 103.

93. Surgeon Thomas Erskine Dempster (1799-1883). Bengal East India Army, 1820-57.

94. Henry Yule (1820-1889). Bengal Engineers, 1840-61. Colonel; K.C.S.I. 1843, assistant on Western Jumna Canal; 1846-54, Executive Engineer, Ganges Canal; 1857-61, Secretary in Public Works Department; 1875-85, Member of Council of India.

95. G.C.W., Vol. 3, pp. 24-58.

96. G.C.W., Vol. 1, pp. 63-4.

97. G.C.W., Vol. 1, pp. 65-6.

98. W.E. Baker, *General instructions for the Executive Officers of the Third, Fourth, Fifth and Sixth Divisions of the Ganges Canal Works*, [1847].

99. See above, Note 68.

100. G.C.W., Vol. 3, pp. 59-110, 158-89.

101. G.C.W., Vol. 2, pp. 549-82.

102. G.C.W., Vol. 2, pp. 544, 547-8.

103. Thomas Login (1823-1874). M.I.C.E. 1868. 1844-74, employed in Public Works Dept., North Western Provinces.

104. Thomas Login, *A Short Description of the Ganges Canal*, Edinburgh 1857, pp. 11-12. Reprinted from *Trans. Royal Scottish Society of Arts*, Vol. 5.

105. G.C.W., Vol. 2, p. 549 ff.

106. Login, *Description of Ganges Canal*, p. 18, Note 104.

107. G.C.W., Vol, 2, pp. 326-8; *Calcutta Review*, Vol. 12, 1853, pp. 502-4.

108. Thomason Civil Engineering College, Roorkee, *Prospectus 1867*, p. 19.

109. P.T. Cautley, *Slope Tables for the Use of the Canal Department, North Western Provinces* ..., Umballa 1851; and P.T. Cautley, ed., *Useful Tables for the Canal Department, North Western Provinces*, [Agra 1851].

110. G.C.W., Vol. 1, pp. 90-1, Vol. 3, p. 10.

111. Login, *Description of Ganges Canal*, p. 12, Note 104.

112. See above, Note 66, Second section, p. 3.

113. Reinhard Woltmann (1757-1837) had invented a current-meter *c*. 1790, but it is hard to know how available such instruments would have been in India in 1840.

114. Pierre Louis Georges Dubuat (1738-1809) had published *Principes d'hydraulique verifie par un grand nombre d'experences, faites par ordre Gouvernment*, Paris 1779; enlarged 1786; posthumous edition, 1816.

115. See above, Note 66, Section section, Appendices IV and V.

116. *G.C.W.*, Vol. 1, p. 197, Vol. 2, p. 538 ff.

117. *G.C.W.*, Vol. 3, pp. 154-7.

118. See above, Note 66, p. 9.

119. Baron Gaspard Riche de Prony (1755-1839). *Recherches physicomathématiques sur la Theorie des Eaux courantes*, Paris 1804.

120. *G.C.W.*, Vol. 1, p. 243, and Vol. 3, pp. 142-6.

121. J.G. Medley, *Irrigation Works*, Thomason Civil Engineering College Manuals, No. 10, second edition, Roorkee 1873, p. 29.

122. James Crofton (1826-1908). Bengal Engineers, 1843-82. Lieutenant-General. 1850-59, assistant on the Bari Doab Canal; 1863-64, Superintending Engineer, Eastern Jumna Canal; 1865-74, Irrigation works, Punjab; 1874-82, Inspector-General of Irrigation and Deputy Secretary to the Government of India.

123. James Crofton, *Report on the Ganges Canal, dated 23rd November 1864*, [1864], p. 19.

124. W. Jeffreys, *Report on the Project for a Canal, from the Ganges, to be termed the Lower Ganges Canal*, Meerut 1870, p. 54.

125. Medley, *Irrigation Works*, pp. 57-61, Note 121; Crofton, *Report on Ganges Canal*, pp. 20-1, 28, 30-2, Note 123.

126. See above, Note 103.

127. Login, *Description of Ganges Canal*, pp. 5-6, Note 104.

128. *G.C.W.*, Vol. 2, pp. 484-5.

129. *G.C.W.*, Vol. 2, pp. 503-8.

130. *G.C.W.*, Vol. 2, pp. 434-6, 453-5.

131. *G.C.W.*, Vol. 3, p. 193.

132. [R. Baird Smith], *Short Account of the Ganges Canal* in English, Hindu and Urdu, [1854], p. 23.

133. *The Delhi Gazette*, Supplement, 12 April 1854, and *The Mofussilite*, Agra, 13 April 1854.

134. John Russell Colvin (1807-1857). Lieutenant-Governor of the North-West Provinces, 1853-57.

135. *Delhi Gazette*, 6 May 1854, 20 May 1854; *Bengal Hurkaru*, Calcutta, 1 May 1854, 15 May 1854.

136. Governor-General's Dispatch No. 487, 11 May 1854, *Calcutta Gazette*, 13 May 1854, pp. 491-2.

137. Minutes of the Council of India, 13 January 1859. India Office Library Records, C/2.

138. See above, Note 35.

139. R. Baird Smith, *The Special Commissioner's Final Report on the Famine of 1860-61*, Calcutta 1861, pp. 32-6. (Supplement to *The Englishman*, 9 October, 1861.)

140. Arthur Thomas Cotton (1803-1899). Madras Engineers, 1820-76. General; K.C.B.; K.C.S.I. 1834-36, 1846, Cauvery Delta Works; 1847-52, Godavari Delta Works.

141. A. Cotton, *Private Memorandum upon the Ganges Canal*, printed as an Appendix in P.T. Cautley, *A Reply to Statements made by Major-General Sir Arthur Cotton, on the projection of the Ganges Canal Works*, London 1863.

142. A. Cotton, *On Irrigation and Navigation in Connection with the Finances of India*, London 1863.

143. Cautley, *Reply to Statements by Cotton*, Note 141.

144. Cotton, *Private Memorandum*, pp. 44-5, Note 141.

145. Cautley, *Reply to Statements by Cotton*, p. 23, Note 141.

146. Ibid., p. 18.

147. Ibid., p. 17.

148. Ibid., p. 18.

149. A. Cotton and P.T. Cautley, *A Discussion, regarding the projection and present*

state of the Ganges Canal, and the measures required to make it reliably useful and profitable, London, 1864.

150. A. Cotton, *Observations on the foregoing Reply* in Cotton and Cautley, *Discussion*, p. 85, Note 149.

151. P.T. Cautley, *A Disquisition on the heads of the Ganges and Jumna Canals, North-Western Provinces, in reply to Strictures by Major-General Sir Arthur Cotton*, London, 1864.

152. Ibid., p. iv.

153. Ibid., pp. 55-6.

154. Ibid., pp. 79, 80-1.

155. Cotton, *Observations on Reply*, pp. 119-20, Note 150.

156. Cautley, *Disquisition on heads of Ganges and Jumna Canals*, p. 88, Note 151.

157. Ibid., pp. 98-9.

158. A. Cotton, *Reply to Colonel Sir Proby Cautley's 'Disquisition on the Ganges Canal,'* London 1864, p. 22.

159. P.T. Cautley, *A Valedictory Note to Major-General Sir Arthur Cotton, respecting the Ganges Canal, with a postscript touching certain misrepresentations of a writer in the 'Times' on the same subject*, London 1864.

160. *The Times*, 2 November 1864.

161. A. Cotton, *Reply to Sir Proby Cautley's Valedictory Note on the Ganges Canal*, London 1865.

162. *The Times*, 19 April 1865.

163. See above, Note 122.

164. Crofton, *Report on Ganges Canal*, Note 123.

165. *The Times*, 26 October 1865, 11 November 1865.

166. Collections to Public Works Despatches to India. India Office Library, L/PWD/3/350, Despatch No. 34.

167. Edward Lawford, Madras Engineers, 1825-71. Colonel. 1861-66, Chief Engineer, Mysore.

168. *Report of the Ganges Canal Committee ...*, Roorkee, 1867; *The Times*, 4 September 1866.

169. India Office Library, MSS Eur. A. 38.

170. Ibid., letter of 26 October 1865.

171. Ibid., letter of 26 June 1865.

172. A. Cotton, *Letter to the Under-Secretary of State for India, on the Report of the Ganges Canal Committee ...*, London 1867.

173. R. Baird Smith, *The Cauvery, Kistnah and Godavery, being a report on the works constructed on these rivers ...*, London 1856, p. 13.

174. H.R. Nevill, *Saharanpur, A Gazetteer*, Vol. 2, of *The District Gazetteers of the United Provinces of Agra and Oudh*, Allahabad 1909, pp. 59-60. See also Morton, *Eastern Jumna Canal*, Note 41.

175. Nevill, *Saharanpur*, p. 65, Note 174; and Central Board of Irrigation and Power, *Irrigation in India*, pp. 64-6, Note 11.

176. George Kuriyan, *Hydro-electric power in India — A Geographical Analysis*, Monograph No. 1, The Indian Geographical Society, Madras 1945. This describes the first seven power-houses; the eighth was added in 1956.

177. Newhouse, Ionides and Lacey, *Irrigation in ... India*, pp. 54-6, Note 2.

178. Sir William Willcocks (1852-1932). 1872-83, Indian Public Works Dept.; 1883-97, Egyptian Public Works Dept.; 1898, Aswan Dam; 1905-11, irrigation works in Mesopotamia.

179. See Medley, *Irrigation Works*, Note 121; and *Professional Papers on Indian Engineering*, first series, Vols 1-7, Rookee 1864-70, second series, Vols 1-10, Roorkee 1872-81, third series, Vols 1-4, Roorkee 1882-85.

180. British Library, Add. MSS 28,599, f. 16. Letter of 13 March 1839.

181. Bust in Town Hall, Calcutta; portrait in Roorkee University.

182. R.M. Panjabi, 'Cautley and the Ganges Canal', *The Geographical Magazine*, Vol. 30, August 1957, p. 169.

183. R. Baird Smith, *Italian Irrigation*, Vol. 1, p. 395, Note 47.

Glossary of Indian names

Modern spelling is given in brackets.

Allumpoor (Alampur)
Beejapur (Bijaipur)
Behut (Behat)
Cawnpore (Kānpur)
Dhunowri (Dhanaurī)
Furruckabad (Farrukhābād)
Fyzabad (Faizābād)
Ganges (Ganga)
Hindun (Hindan)
Hurdwar (Hardwar)
Jumna (Yamuna)
Kulsea (Akalsia)
Kunkul (Kankhal)
Kurnal (Karnāl)
Kuttha Puthur (Katapathar)
Mahewur (Mahewar)
Munglour (Manglaur)
Muskurra (Maskara)
Myapoor (Māyāpur)
Nanoon (Nānu)
Nyashur (Naushahra Tatārpur)
Peeran Kulleeur (Piran Kaliār)
Puttri (Pathri)
Ranipoor (Rānīpur)
Rutmoo (Rātmau)
Sewalik (Siwalik)
Sookurtal (Shokartar Bāngar)
Surkurri (Sarkarī)

On Knowing, and Knowing how to . . .

A. RUPERT HALL

'Logic, as it is generally understood, is the organ with which we philosophize. But just as it may be possible for a craftsman to excel in making organs and yet not know how to play them, so one might be a great logician and yet still be inexpert in making use of logic . . .' So Galileo wrote in his *Dialogue concerning the two chief world systems: Ptolemaic and Copernican*.[1]

One might reasonably suppose that any critical discussion of the writing of the history of technology would require some precise conception of what it is to 'know' or 'to have knowledge of' a subject or technique. But to avoid ambiguities, let us first attempt to be precise about the word 'technology' itself, of which a dictionary definition reads: 1. a discourse or treatise on an art or arts; the scientific study of the practical or industrial arts; 2. by transference, the practical arts collectively. This definition suggests a basic duality of meaning. Either technology is an all-embracing word — the sum of arts and crafts — whose history extends back in time to the neolithic; or technology, like the French *technologie*, implies a high degree of intellectual sophistication applied to the arts and crafts, such as has perhaps only existed within the last few hundred years or at any rate could not have existed before the invention of writing. By speaking of arts and crafts (*arts et métiers*), our ancestors avoided the difficulty; but to write 'a history of the crafts' now would seem distinctly odd. The word 'technics', employed by Lewis Mumford, is not natural in English, nor does his usage have the same meaning as the common cognate form 'techniques' as in the sentence, 'the violinist's technique was superb'. Hence in English the word 'technology' has more and more been generalized to include both the arts and crafts and *technologie*. It was to make this clear that the editors of the Oxford *History of Technology* chose to identify technology with 'making and doing' things; it seemed as broad and as time-free a distinction as could easily be proposed.[2]

Therefore the history of technology is the history of *homo faber*, of man the maker and doer; we can begin this history with the first known artefacts, though it may have begun earlier, and carry it continuously forward to the nuclear age. So much is commonplace.

It is obviously easier, though not easy, to say what has been done or made than how it was done or made. For the whole period of prehistory the how of technology must rest upon inference and analogy; even after written records start, the bulk of material bearing upon technological matters long remains slight and insufficient. Though the historian will always examine attentively the records of technological processes and methods available from antiquity to modern times, he must still rely heavily for his understanding on what he can deduce either from the actual objects or materials produced by these methods or from the workshops, tools and so forth used in their production. The deduction of methods by study of workshops and tools rests very much upon knowledge of the way such similar tools as furnaces and moulds have been used in more modern technologies that are better known to us; one cannot of course be sure that an ancient smith would tackle a certain problem of forging a piece of iron as would a more modern smith, an element of uncertainty that has to be accepted. However, the postulation of ancient technical methods by analogy either with modern practices or with those of modern primitive peoples can be rendered more plausible both by an actual reconstruction of the method — for example iron ore may be smelted in a reconstructed Roman furnace — and by precise analytical examination of historical objects themselves, from which the mechanical processes, heat treatments and so forth to which they have been submitted can often be determined.

Clearly the historical records concerning technology become richer and richer through time so that by the nineteenth century elaborate reconstructions and analyses become less and less necessary; a competent engineer can tell whether a given bridge is made of cast-iron sections or wrought-iron plates without a metallographic examination of the metal, though that might still have its value. Yet even in the nineteenth century there is still much to be learnt about the details of the how of technology by those same methods that are sometimes the sole ones available for earlier periods.

However the historian is often led by the nature of the material evidence as well as by written accounts to ask still more intricate questions. To some the answers are almost trivially obvious. If we ask why the early builders of Western Asia in Mesopotamia and the Indus Valley used bitumen as a mortar and a plaster,[3] it is plausible to see the answer in the water-resisting quality of the material. If asked to explain the occurrence of carvel-built boats in one area of Europe and clinker-built boats in another, the historian postulates the continuance of two independent traditions of boat-building.[4] But when we consider design in boats or structures, this kind of question about the mental processes of the ancient builder or technologist becomes pressing and difficult to answer.[5] I hope it is unnecessary to argue the point that the construction of a Roman aqueduct, for example, required a degree of planning and preliminary organization that was not needed by a Roman potter before he made a pot. The quantity of water required had to be estimated; the quantity and quality of water available from springs or streams estimated; the route surveyed; the methods of overcoming natural obstacles and delivering the water at the point of supply at the proper head established; the size and slope of the channel proportioned to the quantity of water required; and so forth, even before the organization of materials and labour began. We may suppose that Roman engineers made mistakes, but it is hardly likely that they built aqueducts or roads by chance or routine. Because we do possess literary texts, including the relatively ample works of Vitruvius and Frontinus, we can reconstruct the Roman methods of design to a certain extent and even state that certain ideas or principles of design that the Romans held were erroneous, which without the literary evidence we could hardly do.

Here then is a case where literary evidence leads us to the thought of the ancient technologist, which only the existence of literary evidence could do. The historian might establish from measurements that the Romans habitually gave their aqueduct channels an inadequate fall, but the Roman ideas of fluid-flow can only be presented to us in abstract terms and in the absence of writing cannot be reconstructed from the physical remains. The same may be said of medieval cathedrals. While comparative investigation of the surviving buildings can provide much information about the building methods used, the planning of the heights of arches, the diameter of pillars, the provision of flying buttresses and so forth, historians have also learnt a great deal about the methods of design, that is to say the thoughts of the builders, from documentary records.

In the history of technology, I believe the following definitions may be offered: a first-order conjecture implies the reconstruction of a technique based on inference by an analogy or analysis or both, and a second-order conjecture means the attribution of a purpose, design principle or theory to the user of a technique in the absence of literary evidence. Thus a first order conjecture is the outline reconstruction of the methods which were used by Greek or Etruscan potters to produce decorated vessels employing contrasts between two or three colours; there is a complete absence of direct evidence for the firing methods used and for any rationale that may have governed them.[6] But a second order conjecture is our explanation of the existence of welded blades having steely edges and softer central portions when we attribute to the smith a realization that while sharpened edges must be hard, the blade as a whole must not be brittle; in other words, we are putting into the smith's mind ideas of hardness, softness and flexibility for which we have no direct evidence.

As my terminology indicates, second order conjectures involving purpose, design and abstract concepts must necessarily be far more frail than first-order conjectures. In the example of the composite blade weapon, it might alternatively be supposed that this method of forging a blade was worked out by an intuitive evolutionary process involving (like all discoveries properly so termed) an element of chance. It is not essential that we attribute to the ancient smith abstract ideas of hardness, softness and flexibility and an awareness of an association between these ideas and pieces of metal prepared in given ways. One might rather postulate that the smith was as ignorant of such abstractions as of the metallographic structure of the various pieces of iron and steel. But my sole concern at this point is to argue that if we do wish to assign such abstract ideas to the ancient smith — and equally to all other craftsmen — we do so in a highly conjectural, almost poetic, fashion. It would certainly be a very great error to attribute to ancient technologists any abstract ideas for which we have no warrant in early technical literature. For example, we know that the oldest traditions and literature of metallurgy attach a mysterious importance to the nature of the fluid in which hot steel is quenched and recommend more than the use of olive oil or water for the purpose. We may therefore reasonably suppose that the most ancient smiths would have had a ritualistic attitude towards this strange and critical technique of hardening.[7] If, on the other hand, our literary evidence showed that simple fluids were invariably used, we would have no evidence for a different attitude among the smiths.

At this point we may return to our initial problem of what is *knowing* in a technological context. The statement 'the ancient smith knew how to forge a composite blade' is consistent with the statement 'the ancient smith had no metallurgical knowledge'. These statements are strictly similar to the single sentence, 'that boy knows how to ride a bicycle skilfully but he is ignorant of mechanics', which occasions no surprise because we do not expect a bicycle rider to understand mechanics or human physiology. Knowing how to ride a bicycle consists in acquiring a certain set of neuro-muscular co-ordinations as does mastery of many technical skills such as throwing a clay pot, making a good saw cut and ploughing a straight furrow. In the old days the principal instructor in such things was the clip over the ear. But clearly craft consists of more than acquired neuro-muscular co-ordination; it involves experience, a trained eye and a proper sensitivity to the fitness of things that may almost be called aesthetic, and in certain situations even demands an intellectual knowledge that is distinct in kind from the 'knowing how to' used here. But when we say 'that girl knows how to play chess' we imply far more than an acquired co-ordination or even memory of the pieces, board and rules, for someone who knows how to play chess must at least have knowledge of the simplest strategies. But here by 'knowing how to' I mean mainly non-cerebral knowledge as in bicycling, swimming, dancing and other familiar non-craft activities.

Clearly the spectrum of knowledge from knowing how to intellectual or theoretical knowledge is very great. Let us elaborate Galileo's example of the spectrum associated with the organ. Its construction may begin with craftsmen who know how to plane wood, tan skins, smelt metals and so forth. If we transport ourselves back only a few centuries in time, their essential skills could neither be analysed nor even described in a meaningful way, any more than one can learn to ride a bicycle or swim by reading a book. However, the actual organ-builder, though he may be unable to play the instrument and be wholly ignorant of musical theory, must know what an organ is and all its parts in their appropriate sizes, assembly, fitting, rectification and so forth;

he can learn something of his art from books, but abstract knowledge alone would be far less serviceable than practical knowledge alone. Then there is the player; he may even be totally ignorant of the notation of music, but know how to play the organ by ear, which is yet another exercise in neuro-muscular co-ordination of which all performance in part consists. Or the player may be Johann Sebastian Bach. Yet the composer of music does not necessarily have to be, as were Bach, Mozart and Beethoven, a virtuoso performer on an instrument any more than he needs to have a knowledge of the science of acoustics. At the same time the physicist whose science is acoustics need not be able to play or compose or to construct a musical instrument.

Hence the statement, 'Jones has a knowledge of musical instruments', is meaningless by itself. It might mean that Jones is knowledgeable as a particular kind of antiquarian; that he is a good performer on violin, flute and piano; that he keeps a shop selling musical instruments; that he writes music for symphony orchestras; and so forth; the possible meaning could only be learnt from the context. The sentence, 'Jones knows how to cut a whistle', would be more precisely informative, but in turn has no implication for his ability to perform on the whistle, knowledge of harmony and counterpoint and so forth. Suppose we find in some Danish bog a miraculously preserved mesolithic whistle. This is equivalent to the statement, 'Jones knew how to cut a whistle'. Let us also suppose the whistle has finger-stops. We cannot be certain that its maker made music since the stops might have been used for signalling or instructing well-trained sheep-dogs, but let us consider the idea that this mesolithic whistle was used to make music. We still cannot have any notion of what sort of music it was, in what social context it was played and what it meant to player and listeners, and indeed are unable to penetrate in any way into the minds of those remote people from the mere discovery of this artefact, interesting as it would be.

So far as history before the last few centuries is concerned, it is futile to demand, as a recent writer has done, a history of technology as knowledge on the ground that it is wrong to 'deny technology a significant component of thought'.[8] It is worse than futile to proclaim such a programme without considering the rather elementary distinction between knowing how to and intellectual knowledge and without looking at the different connotations of the verb 'to think'. The poet may 'think himself transported to the distant fields of home' or I may think myself lucky, or think about Tom Thumb as Dr. Johnson recommended in certain situations, but if we are to introduce thought in a more important way into the history of technology, a more precise definition of the thinking is necessary.

We may confidently believe that from prehistory to the present day craftsmen have pondered not only over their problems and failures, but also over their successes, especially when the latter resulted from variations upon established techniques, whether accidental or not. The problem is that for the reason I have already endeavoured to explain the mental processes of early technologists are largely hidden from us; only documents can reveal them authoritatively, and then but partially. If we try to reconstruct the thoughts of such craftsmen we can only create a mirror reflection of what is in our own minds; even if what is in our own minds is borrowed from Vitruvius or Theophilus we still cannot cease to be ourselves. It is mere self-deception to suppose that one can recreate thought from things, as for example those who have sought to fabricate a palaeolithic religion based upon the evidence of cave-paintings or, perhaps, an iron age astronomy on the evidence of stone circles. In such cases we do not truly see our forbears, but only see an image of ourselves distorted according to our own fancies and unconscious urges.

As for knowing how to, this is a matter of fact or at any rate of first-order conjecture. We have learnt that the Sumerians knew how to solder metals; we infer that they also knew how to anneal metal as when raising a bowl with the hammer, for this seems implicit in their successful technique.[9] Indeed, all the ancient peoples knew how to do many things by methods which are far from clear to historians, though plausible hypotheses can be suggested, for their alignment of tunnels and roads over quite long distances.[10] Our ignorance arises not so much from inability to imagine any simple method by which the ancient technologist might have solved the problem as from inability to choose between a number of possible methods. Thus, at one level, the statement of the achievement of early technology is nearly equivalent to a statement of the knowledge of the ancient technologist: we say that he knew how to do this because he did it, which is, in fact, our contemporary test of deciding whether Brown knows how to do something. If we seek after some higher levels of cerebral activity, we may simply be deluding ourselves, either because knowing how is essentially a matter of acquired muscular co-ordination and experience or because we simply have no evidence to go on. This last point is particularly impressive in considering some of the pivotal technical innovations of the past. Let us take the water-mill and windmill, where the former was introduced into Western Europe over a thousand years earlier than the latter. We have a fair amount of information on the early history of both devices; we have Vitruvius's text concerning the water-mill; we can compare the windmill with its Asian precursor and with the watermill; but our information is far too slender to permit us to make even the least guess at the mechanical knowledge, design or experiments of the first inventor or inventors in either case. And indeed we are hardly better off with Newcomen's steam-engine.

What then of knowledge in technology which is other than knowing how? Firstly, without pressing the evidence too hard we may claim that some such knowledge is at least as old as the civilizations of Mesopotamia and Egypt, whose documents make it manifest to us, but it must have been a tiny constituent in all early technological activity, which was after all mainly primitive farming. Otto Neugebauer has written, 'one must simply realize that mathematics and astronomy had practically no effect on the realities of life in the ancient civilizations . . . [And in particular] the requirements for the applicability of mathematics to problems of engineering are such that ancient mathematics fell far short of any practical application'.[11] His judgement is capable of wider extension; the higher learning represented by writings on chemical topics for example must also have been of very limited application.[12] Nevertheless, the surviving technical literature grows richer through the Graeco-Roman period and then lapses for several centuries to revive again in both Islam and Western Europe; incidentally, I do not here consider the Far East. Secondly, it may be remarked that those who knew often also knew how; whatever the status of Vitruvius may have been, and Frontinus was certainly an administrator, Hero like al-Jazarī or Theophilus later surely wrote of techniques with which he was familiar in practice.[13] Thirdly, we must recognize that a fair proportion of all documentary material in technology, up to the Renaissance and later is concerned with design as an aesthetic, which does not concern me here, and design as a theory of proportion, for things are made of different sizes and rules have to be made accordingly.

Now the concept of rule is in this context extremely important for we are not here dealing with rules of technical practice but with an abstraction. A craftsman might thoroughly know how to build a ship without awareness of such rules, but knowledge of them would enable him to design whole series of ships. Moreover, such rules were not dictated by the immediate requirements of technique as are the technical rules — if the technical rules are ignored, soldered joints will fall apart, corners will not be square and so forth — but represent a rationale of technique corresponding to theory in scientific technology. However, we should be careful not to over-emphasize the significance of canons of design because of the outstanding interest of their appearance in our documentary sources; they are always known in a constructional context as in Vitruvius's *De Architectura* and later on in manuals of building and ship construction, but were never relevant to most ordinary trades or tradesmen. Actually a great part of all the technical literature before the eighteenth century is concerned with technical rules, recipes, gadgets, 'secrets', and of course pleas for the patronage of the great and wealthy. We should assign such technical literature to the realm of knowing how rather than to what we are seeking for here, that kind of higher knowledge which caused the word 'mystery' to acquire its current meaning.

Now it may justly be argued that the extant technological literature before 1700 does not give a completely fair picture of the character of prescientific technology, and moreover that this literature has never been systematically analysed in an attempt to trace the history of technology as knowledge rather than as knowing how. Students of Leonardo da Vinci would surely wish to emphasize the high sophistication of his technological studies as in the movement of water and his endeavour to marry technology and philosophy, and it might be argued that he can hardly be a unique figure. On the other hand, however, it is obvious that even the best technical writing of the sixteenth century such as that by Agricola and Biringuccio is practical and descriptive in character and rather notably deficient in generalizations and explanations. Dr. Alex Keller's study of the machine-books hardly alters this general picture;[14] their authors are wholly concerned with the particular and 'practical' (though really highly fanciful), pursuing their mechanical hobby-horses *ad nauseam* through endless permutations and demonstrating repeatedly their total failure to grasp the most elementary theoretical notions about the functioning of machines such as the desirability of minimizing friction. In fact, the overt literature of technology in the sixteenth century is very far from reflecting accurately the success of craftsmen in knowing how to achieve all kinds of impressive and delightful new results. The growth in size of ships and the power of their armaments, the increasing refinement of furniture-making, ceramics, textiles and other consumer crafts; the application of coal to industrial processes; the ever-extending uses of wind- and water-power; all these constitute but a few examples of growing technological competence which is still, as in antiquity, much better-known to us from extant objects or from incidental records than from self-conscious descriptive treatises. Agricola's *De re metallica* naturally reflects the mining progress of the past; no successor described the further stupendous achievement of the next one and a half centuries.

If the increasing scope and refinement of the knowing-how of technology has to be painfully traced by studying multifarious historical sources, where is the growth of the knowledge of technology to be found? That such knowledge does grow is surely beyond doubt: it is to be found in the *Mémoires* of the Académie Royale des Sciences, in the *Philosophical Transactions* and Smeaton's *Reports*, in the encyclopaedias, in the writings of French and British engineers, in the *Transactions* of local societies, in the curricula of technical schools and eventually in professional examinations. Long before the end of the nineteenth century a huge body of knowledge, much of it mathematical in form, had come into existence relating to the design and performance of all kinds of bridges, steam-engines, boilers, heavy electrical engineering equipment, submarine cables, chemical works and even gliders and aeroplanes. Here surely we can see at last technological knowledge (*technologie*) clearly separate in character from the age-old knowing how to, though of course, they are related at the industrial level. Whence did this new knowledge arise? Nearly all historians have regarded it as deriving more or less directly from science; some have regarded it as essentially applied science;[15] and others, probably more correctly, as partly applied and partly imitated science. For often the engineer found scientific data inadquate and scientific theories too cumbersome; therefore he had to institute experiments to complete the data and introduce simplifications to make the theories usable. But there are obvious and well-known reasons for regarding whole new industries such as the electrical industry as science-based, just as there are equally good reasons for regarding the new prime movers devised rather later by Diesel and Parsons as the fruit of sophisticated scientific ideas which ultimately originated in the thermodynamic concepts of the 1850s. As L.J. Henderson reputedly declared in a memorable phrase it is not in dispute that science owed a large debt to the heat-engine, but in the second half of the nineteenth century that debt was more than repaid. The evolution of the new energy converters proceeded in a very different way from that of the old Cornish engine, now gradually falling into disuse.[16] In short, if one is to characterize the differences between the level and kind of technology practised around 1820 and that practised around 1890, virtually every historian would single out the origin of new technological ideas in scientific work and the training of technologists in mathematics and science, so that they were equipped to apply both analytical and experimental skills in their own work. It is a commonplace of educational history that, in order to produce qualified engineers of this new type, the structure of technical education was remodelled and extended in most civilized countries, on the widely praised German model of technical educations.

Many students of the Industrial Revolution in Great Britain have supposed this transforming influence of science upon technology to begin as far back as the first half of the eighteenth century, supposing this influence to have been made fruitful, in turn, by the Scientific Revolution of the sixteenth and seventeenth centuries.[17] These historians regard scientific method and scientific knowledge as major contributors to the technological innovations that both preceded and with increasing frequency accompanied the industrial revolution. For fear of being misunderstood, let me at once add that these historians do not exclude the role of empirical inventiveness, nor do they overlook the importance of wholly non-technical, economic and social factors in the British 'take-off'. I have been a sceptical critic of this interpretation, although I have not disputed the view that the increasingly scientific character of technology in the nineteenth century signalizes a turning-point in the history of material civilization. Indeed, I also agree that much scientific investigation of a useful or applied character was carried out in the eighteenth century, notably by the French school of mathematicians, culminating in the work of Lazare Carnot and Coulomb.[18] What I doubt is the validity of such investigation as an explanation of the Industrial Revolution in Britain until well after 1800.[19]

Naturally, just as in the eighteenth century there were eminent technologists like Smeaton, Watt, Wedgwood and other members of the Lunar Society with strong scientific connections — men to whom some would perhaps think the label 'scientist', though anachronistic, fully appropriate — so in the next century the more scientific technology professed by a Wheatstone or a Siemens, a Parsons or a Marconi was by no means devoid of important empirical elements. Thus Marconi had to learn by experiment that radio waves did not necessarily possess optical properties, that antennae of certain shapes revealed marked directional properties, that short waves were reflected by the ionosphere and so forth. Further, since technology deals with men and materials, neither of which are constant in their behaviour, the rules and tables used by engineers often codified the results of experience rather than experiment or theory. I do not think anyone would maintain, therefore, that an engineering discipline based on physical science and mathematics which contained major elements of analysis and theory is the same as a department of physical science.

One recent author has departed strongly from the general consensus that the penetration of science into technology is a historical reality. Dr. Layton argues that 'if basic science is the source of all new technical knowledge, then technology itself produces no new knowledge, and the technologist's role becomes that of applying knowledge generated elsewhere'.[20] To declare that technology produced 'no new knowledge' in the nineteenth century would certainly seem to be an exaggeration; and perhaps no one has ever made such an extremely negative claim. What may properly be said is that basic knowledge tended increasingly to derive from science, while knowing how to apply it to useful ends was very much the province of the technologist. Can any one doubt that knowledge of electromagnetic phenomena came from Oersted and Faraday, while the knowing how to use it was built up by Wheatstone and a host of other telegraph inventors and engineers? Or that electromagnetic waves were discovered by Hertz while Marconi was the one person in his age who knew how to make them useful? The development of wireless telegraphy in fact serves as a discriminating instance of what may happen generally: as the development of the new technique proceeded, so further injections of scientific knowledge by such people as Lodge and Fleming enriched it, while the technique itself developed its own logic and experience so that it was certainly distinguishable from university physics.

To recognize this distinction between physics and electrical engineering, which adds experience and knowing how to a part of the scientific content of physics, is not to claim, as Dr. Layton seems to wish to do, that science and technology are separate bodies of thought differing from each other in significant ways. In my view, it is in basic ideas that science and modern technology most closely approximate to each other. Dr. Layton enlists in his support, perhaps incautiously, the testimony of Aristotle and Alexandre Koyré, whose point he quotes[21] and seems to misunderstand. Koyré argued that technology was bad science. Moreover he held, as is obviously true, that the rigour and exactitude of science do not belong to the world of technology, but even this statement is less true today than it was in 1948 when, for example, neither industrial nuclear power nor commercial supersonic flight had yet been attained. Koyré contrasted science's world of precision with the engineering world of 'near enough'. But it seems an exaggeration to claim that imprecision and compromise, inevitable to engineering design as they may be, constitute another world of thought. If a physicist expresses dimensions in ångströms while an engineer is content with hundredths of a millimetre, the velocity of a radio wave, for example, is the same for both as are the basic formulae and parameters.

One may reasonably assert, I think, that in broad terms science in the nineteenth century, and to a lesser extent earlier, contributed to technology: 1. mathematical analysis, extended from physical science to engineering; 2. the method of establishing facts by carefully controlled experiments; 3. knowledge of relevant natural laws such as those of thermodynamics or genetics; 4. acquaintance with new natural phenomena such as electromagnetism or catalysis. These contributions at least came to technology from science, and I cannot conceive of late nineteenth-century technology without them; accordingly, I think the areas of technology and industry into which these contributions entered are very different from areas lacking such transforming injections of new ideas and methods as in the wrought-iron industry. Knowledge of making wrought iron in a puddling furnace consists entirely in doing it; a man either knows how to do that hot and exhausting job at the furnace mouth with his own hands or he does not; no literary description, formulae or pyrometric measurements can ever be substituted for experience of the look and feel of the metal. Some element of this same experience, skill and familiarity with the feel and look of things may remain in all levels of technology, but gradually loses significance as the formulae, measurements, print-outs and so forth come in. The technologist is, moreover, always concerned with doing and making and is not concerned like the scientist with experimental and conceptual problems. It is not the same thing to have knowledge of the theory of structures as to know how to build an aeroplane fuselage.

Thus, there is admittedly a sense — though a very different sense from that prevailing in Aristotle's time — in which the distinction between knowing and knowing how to is still valid so far as technology is concerned; and the knowing part has come to be almost wholly scientific in content and character. If this were all Dr. Layton's claim I should accept it, but he claims more by saying that there are separate bodies of knowledge.[22] This seems to me wrong because there are overlapping bodies of knowledge. To a very great extent technologists and scientists are on common ground; the scientist moves from the common ground into an area of refinement, complexities and new research that the engineer does not need at the moment, while the engineer moves from the common ground into a world of drafting, computations, approximations and compromises between conflicting optima — not to mention finance — into which the scientist need not enter. The 'laws of science refer to nature and the rules of technology refer to human artifice',[23] writes Dr. Layton, but he apparently forgets that the 'rules of technology' can only be valid in so far as they are consistent with the laws of nature and that these 'rules' have been continuously adapted during the last few centuries to make full use of science's increasing understanding of nature's laws. Since Dr. Layton quotes Aristotle with approval, let me in return refute him with a quotation from that great critic of Aristotle, Galileo:

> It would be novel indeed if computations and ratios made in abstract numbers should not thereafter correspond to concrete gold and silver coins and merchandise. Do you know what does happen, Simplicio? Just as the computer who wants his calculations to deal with sugar, silk and wool must discount the boxes, bales and other packings, so the mathematical scientist, when he wants to recognize in the concrete the effects which he has proved in the abstract, must deduct the material hindrances, and if he is able to do so, I assure you that things are in no less agreement than arithmetical computations. The errors, then, be not in the abstractness in concreteness, not in geometry or physics, but in a calculator who does not know how to make a true accounting.[24]

The history of modern technology has been in part the mathematical analysis of the 'material hindrances' which do indeed cause technological problems to resist the over-simple theories which science before the nineteenth century was alone capable of producing.

Notes

1. Galileo Galilei, *Dialogue concerning the two chief world systems: Ptolemaic and Copernican*, trans. Stillman Drake, Berkeley and Los Angeles, Calif., 1953, p. 35.
2. See also Charles Singer, 'The happy scholar', *Newcomen Society transactions*, 29 (1953-5), 123-32 (Dickinson Memorial Lecture). It should hardly be necessary to add that in giving technology a universal connotation as all ways of making and doing there was no intention to correlate a technology with manual labour; for example, one way of making things is to use a computer in designing a bridge.
3. See for example C. Singer and others (eds.), *A history of technology*, 5 vols., Oxford, 1954-8, vol. 1, pp. 250-6 (hereafter cited as *Oxford history of technology*).
4. *Oxford history of technology*, vol. 2, p. 578.
5. See A. Rupert Hall, 'The genesis of engineering: theory and practice', *Society of Engineers journal*, 60 (1969), 17-31, which attempts to make a small contribution to the history of design.
6. *Oxford history of technology*, vol. 2, pp. 262-6.
7. *Oxford history of technology*, vol. 3, p. 35.
8. Edwin T. Layton Jr., 'Technology as knowledge', *Technology and culture*, 15 (1974), 31-41, p. 32.
9. *Oxford history of technology*, vol. 1, pp. 624-5, 636-7, 649-54.
10. June Goodfield and Stephen Toulmin, 'How was the tunnel of Eupalinus aligned?', *Isis*, 56 (1965), 46-55.
11. Otto Neugebauer, *The exact sciences in antiquity*, Copenhagen and Providence, N.J. 1951, pp. 71-2.
12. I have particularly in mind the mass of cuneiform material dealing with chemical arts; see J.R. Partington, *Origins and development of applied chemistry*, London, 1935 and the many articles published during the last twenty years by Dr. Martin Levey (*Isis*, 47 (1956), 287 lists seven early papers by him). Despite this wealth of evidence, it is impossible to believe that in a predominantly illiterate society the transmission of technical information by written texts can ever have been of massive importance.

13. Hero of Alexandria, *The pneumatics*, 1851, trans. Bennet Woodcroft, facsimile reprint ed. Marie Boas Hall, London, 1971; Al-Jazarī, *The book of knowledge of ingenious mechanical devices*, trans. Donald R. Hill, Dordrecht and Boston, Mass., 1974; Theophilus (called also Rugerus), *The various arts*, trans. C.R. Dodwell, London, 1961 and *On divers arts: the treatise of Theophilus*, trans. John G. Hawthorne and Cyril Stanley Smith, Chicago and London, 1963.

14. A.G. Keller, 'Early printed books of machines, 1569-1629', University of Cambridge unpublished Ph.D. thesis, 1967. Dr. Keller's broad conclusions are apparent in his *Theatre of machines*, London, 1964.

15. The words 'applied science' have proved unfortunate as conveying the idea that science provides a recipe or prescription for some device or process, which the technologist has only to realize on the manufacturing scale; it is universally recognized that this provides a highly unrealistic model of the relations between science and technology. However, the original emphasis of the expression was probably rather to draw attention to the application of scientific methods and attitudes in technology; in this sense Smeaton's investigations of the efficiency of wind- and water-mills were certainly a work of applied science (and the results were published by the Royal Society) although Smeaton borrowed no general theories, parameters or calculations from the science of his day in carrying out his enquiries.

16. D.S.L. Cardwell and Richard L. Hills, 'Thermodynamics and practical engineering in the nineteenth century' in A. Rupert Hall and Norman Smith (eds.), *History of technology: first annual volume, 1976*, London, 1976, pp. 1-20.

17. A.E. Musson and Eric Robinson, *Science and technology in the Industrial Revolution*, Manchester, 1969; A.E. Musson (ed.), *Science, technology and economic growth in the eighteenth century*, London, 1972; Arnold Pacey, *The maze of ingenuity: ideas and idealism in the development of technology*, London, 1974.

18. C.C. Gillispie, *Lazare Carnot savant*, Princeton, N.J., 1971; C.S. Gillmor, *Coulomb and the evolution of physics and engineering in eighteenth century France*, Princeton, N.J. 1971; Jacques Heyman, *Coulomb's memoir on statics: an essay in the history of civil engineering*, Cambridge, 1972. For the extraordinary vigour of France in 'useful chemistry' as well as applied mathematics, see Henry Guerlac, 'Some French antecedents of the Chemical Revolution', *Chymia*, 5 (1959), 73-112.

19. See a fuller discussion in A. Rupert Hall, 'What did the Industrial Revolution in Britain owe to science?' in Neil McKendrick (ed.), *Historical perspectives: studies in English thought and society in honour of J.H. Plumb*, London, 1974. It is curious that while such historians as Professor Musson and Dr. Pacey (see note 17 above) have criticized my ideas as expressed in 'Engineering and the Scientific Revolution', *Technology and Culture*, 2 (1961), 333-40, on the ground that I underestimated the role played by science in the development of modern technology, Dr. Layton (see note 8 above) criticizes the very same article for its over-emphasis of the scientific character of modern technology. To satisfy both sets of critics, it seems that I would now have to argue that technology developed by applying science in the eighteenth century, but did not progress in the same way in the nineteenth, which seems absurd.

20. Layton, 'Technology as knowledge', p. 34.

21. Alexandre Koyré, 'Du monde de l'à-peu-près a l'univers de précision', *Critique*, Paris, 4 (1948), 806-23. Compare Layton, p. 36.

22. Layton, 'Technology as knowledge', pp. 36-40.

23. Layton, 'Technology as knowledge', p. 40.

24. Galileo, *The two chief world systems*, trans. Drake, pp. 207-8.

Vitruvius and the Elevated Aqueducts

FRANK D. PRAGER

Vitruvius's *Architecture* is a classic work packed with technical information and full of historical problems. This essay concerns the work's position on the technology and history of elevated aqueducts, which is beset by questions about the aqueducts themselves as well as the aqueduct text.

The ancient aqueducts were magnificent. Their channels had been elevated on boldly constructed arcades, which extended high above the plains, to conduct water flowing by gravity from distant hills into the cities and towns. Even the ruins are still impressive (Fig. 1).[1] In some cities, mainly in Spain, Roman aqueducts still operate. In Rome itself, the Porta Maggiore is still intact and it is possible to visualize how the

Figure 1. Elevated aqueducts: ruins of the Aqua Claudia, with remnants of the Anio Novus on top, a few kilometres south-east of Rome. While many arches were destroyed or used as quarries for building material, some arches (seen here in the middle of the picture) were repaired or supplemented by lower arches.

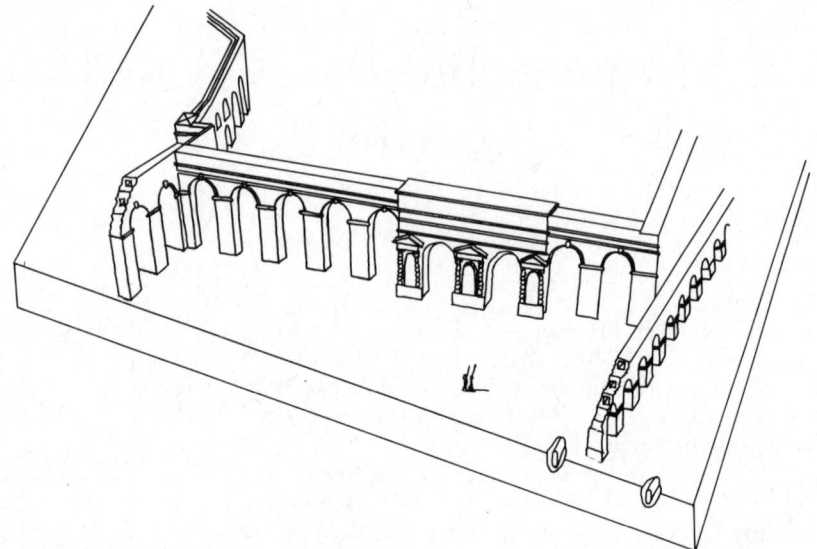

Figure 2. The major aqueducts entering ancient Rome. At right, the pre-Vitruvian ducts: The Anio Vetus below ground, and near it, somewhat above ground, the Aqua Marcia, carrying the Aqua Tepula and then the Aqua Julia. More highly elevated, and extending over the Porta Maggiore, the post-Vitruvian ducts: the Aqua Claudia and the Anio Novus on top.

aqueducts built at different times converged and extended over or near or under its arches (Fig. 2).

It has also been possible over the centuries to envisage in principle the entire course of the ducts (Fig. 3). However, accurate mapping was not easy. Some kilometres outside Rome, the elevated ducts disappear into the ground; conjecture or difficult searching is needed to establish the further course of each duct towards the source-head. Only in the present century was it possible to establish clear facts about this vast underground network (Figs. 4 and 5). This task was promoted mainly by a team of distinguished workers: Ashby, an English expert on Roman architectural history; Reina, an Italian expert on surveying optics; and Van Deman, an American archaeologist of Rome.[2] The results of their teamwork are still in need of study.

Other problems are encountered in the Vitruvian treatise, particularly in the notes on elevated aqueducts.[3] If these notes were originally accompanied by drawings or maps, which is unlikely, the illustrations have vanished long ago.[4] The text was criticized in late antiquity as being too wordy for practitioners: about A.D. 30 the abbreviator Cetius Faventinus called it 'extensive' as well as 'learned' and pointed out the need for a popular abridgement, which he produced — beginning with the aqueduct part — in order 'that more common workers be not prevented by the long and complex verbiage from studying the contents'.[5]

Today, by contrast, the old descriptions seem rather too short and

D uctus autem aquæ fiunt generibus tribus, riuis per canales structiles, aut fistulis plumbeis, seu tubulis fictilibus, quorum eæ rationes sunt. Si canali/ bus, vt structura fiat q̄ solidissima, solumq̃ riui libramenta habeat fastiga/ ta ne minus in centenos pedes semipede, eæq̃ structuræ confornicentur, ut minime sol aquam tangat.

Figure 3. Aqueduct text and picture from *M. Vitruvius per Jocundum solito castigatior factus* ..., Venice, 1511, folio 81r. At *c. fornix*, the covered channel construction.

dry for historical evaluation. According to a recent commentary by Callebat, Vitruvius 'disregards the different data affecting the arcaded aqueducts, which concern their economics ... aesthetics ... and technology ...'; and also shows 'an irritating inability to distinguish the essential from the details'.[6]

We will investigate these difficulties. How does the Vitruvian text relate to the actual elevated aqueducts? How verbose or laconic is it? Why did it confuse the different readers? How did its author try to handle the matters which were essential at his time?

Vitruvius's qualifications as an authority on aqueducts

One of the problems about Vitruvius until recently was his dates. Because of Latin peculiarities encountered in the text, the *Architecture* was sometimes called a medieval production, and its dedication to the Emperor Augustus a forgery. However, reliable evidence shows that the first draft originated about 25 B.C., allowing for a variation of five to ten

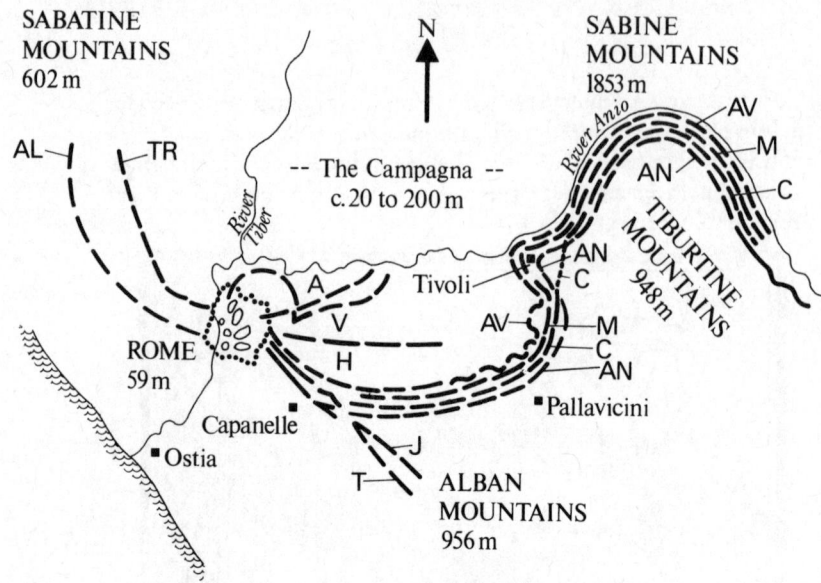

Figure 4. General plan of Rome's aqueducts

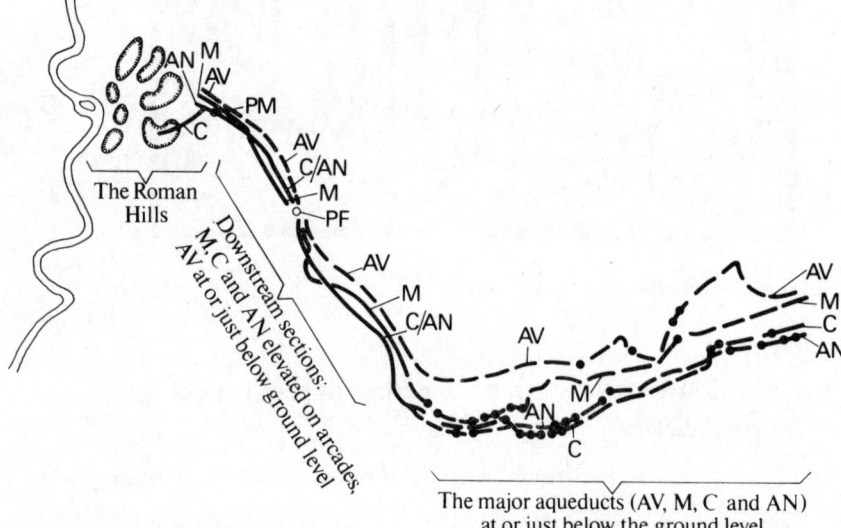

Figure 5. Detailed plan of the major aqueducts near Rome

KEY

A	Aqua Appia (4th century B.C.)	H	Aqua Hadriana (Alexandrina) (2nd/3rd centuries A.D.)
AV	Anio Vetus (3rd century B.C.)	PM	Porta Maggiore on Via Praenestina
M	Aqua Marcia (2nd century B.C.)	PF	Porta Furba on Via Tuscolana
T	Aqua Tepula (2nd century B.C.)	●	*putei* and *cippi* located by T. Ashby
J	Aqua Julia (1st century B.C.)	■	towns
V	Aqua Virgo (1st century B.C.)	———	Elevated on continuous arcades
AL	Aqua Alsietina (1st century A.D.)	– – –	At or just below ground level
C	Aqua Claudia (1st century A.D.)	······	Deeply below ground level
AN	Anio Novus (1st century A.D.)		
TR	Aqua Trajana (2nd century A.D.)		

years either way for preparation and completion. The work is therefore definitely of the Augustan age.⁷

Other problems relating to the technical notes in the work were introduced by subsequent writers who paraphrased, abridged or copied the work and in some cases misunderstood and corrupted the engineering expressions, which were converted into meaningless phrases. In modern times some of these errors were emended by such writers as Fra Giocondo, a Renaissance engineer, and Rose, a philologist of the nineteenth century.⁸

There is no firm evidence that Vitruvius had the qualifications of a water engineer or for teaching such work. He himself makes the pertinent assertions that he introduced a new system of urban water distribution and that he had hydrological experience and book-learning.⁹ As an old man he writes nostalgically about the traditions of the 'ancients' in fields of architecture other than aqueduct building.¹⁰ He does not claim that he invented anything concerned with elevated aqueducts. If he planned or supervised water systems, whether in Italy or the major provinces, the work is likely to have taken place under Agrippa, who was Augustus's son-in-law, counsellor, general and waterworks administrator.¹¹

There are some items of information relating to remote places in the provinces. Vitruvius says he discussed water supply with a nobleman of Numidia;¹² and Algeria has furnished archaeological information which might indicate his presence there. At Thibilis (now Announa), parts of public baths and aqueducts belonging to the aquae Thibilitanae were excavated and inscriptions found.¹³ Two of the inscriptions contain the name Vitruvius as well as the given name Marcus, which is usually assigned to the architect; whereas one includes the name Mamurra, which has been asserted and denied to be the cognomen of Vitruvius.¹⁴

The Vitruvian treatise mentions building traditions in Africa and other regions. It has also been asserted — and denied — that the author came from Numidia.¹⁵ At this point it is pertinent to note two inscriptions, the final one in the paragraph above and another which has hardly been studied in connection with aqueduct arcades.

This Mamurra inscription, which is on a marble plate, is described by the excavator only as having been found in front of 'an edifice with arches' and reads:

M * VITRVVIVS
MAMVRRA
ARCVS
S * P * F

MARCUS VITRUVIUS
MAMURRA
BUILT THESE ARCHES
AT HIS OWN EXPENSE¹⁶

The other plate, which seems to come from the tomb of a woman related to Marcus Vitruvius, is a fragment which reads:

VITRVVI POT . . . .
TVLA * V * A XX . . . . .
M * VITRV . . . .

VITRUVIUS' . . . .
. . . . LIVED . . . YEARS
MARCUS VITRUV . . . .[17]

Together, the two inscriptions show that a man or two men called Marcus Vitruvius, who might not be our architect but might have been related to him, lived in Thibilis; that he, or one of them, financed the building of arches and perhaps technically directed the project; and that this building occurred near the 'waters of Thibilis', at an unknown date in antiquity, conceivably, as part of an aqueduct. Unfortunately, the treatise is silent on these matters.

In post-Vitruvian antiquity our architect or his treatise was mentioned occasionally.[18] We find a positive although not an enthusiastic view in Pliny the Elder's *Natural History*, an encyclopedia dedicated to the emperor Titus in A.D. 77. It mentioned Vitruvius as a scientific and technical authority on 'stones' and is silent about the treatise, but closely paraphrases the aqueduct chapters.[19] We find a somewhat similar view a few decades later in Frontinus's work on aqueducts where 'Vitruvius the architect' is mentioned in connection with the distribution system.[20] Pliny and Frontinus were naturally interested first of all in their own achievements, but both treated Vitruvius as a recognized builder and teacher in the indicated fields.

As already mentioned, a more critical attitude was adopted about A.D. 300 by Faventinus, but the treatise was considered authoritative. Certain sections, especially those on foundations and piers, remained very influential in the Middle Ages, while other sections, principally those on the classic orders, were completely forgotten in the Romanesque and Gothic periods; they found a revival of interest in the Italian Renaissance, but were promptly forgotten again in the Baroque period.[21] Among humanists and for some time among architects in the 1500s, Vitruvius gained enormous fame, and such writers as Cardano admired him as a unique genius of both science and the arts; they blamed the 'obscurities' of his treatise on his copyists.[22]

In the late 1700s, renewed classical studies made it clear that Vitruvius had sometimes been overestimated, and that his significance for Graeco-Roman art and science must be evaluated realistically. Some modern writers may then have over-reacted, declaring his work and most or all of Roman science, to be negligible.[23]

Some writers of the 1800s even charged him with plagiarism. This charge was surprisingly made by Rose, one of those who purified the text of errors introduced by copyists. He noted remarks by Vitruvius which had close parallels in the work of the earlier writer Varro — a source acknowledged in the treatise — and promptly concluded that Vitruvius 'copied everything' from Varro. Another antiquarian, Oder, found this theory exaggerated, but still believed that Vitruvius made false pretences about his sources and that he was 'the typical half-educated proletarian'.[24] Such disdainful views, as well as the former hero-worshipping, should be rejected.

Vitruvius's notes on elevated aqueducts

In his introduction on architectural harmonies, Vitruvius discusses matters to be studied by the architect. He states in Book I chapter i paragraph 7 and 8 that those matters include 'physics . . . of conducting water' as taught by Ctesibius or Archimedes and also 'canons of music'. Shortly thereafter he points out that there is a 'common bond' between these matters.[25] The statement was later paraphrased as suggesting that in aqueducts utility and beauty 'should be combined'.[26]

Perhaps the statement can be paraphrased more clearly in modern terms: there is aesthetic satisfaction in functional perfection which is achieved by such structures as aqueducts. Other paraphrases of course are also possible. It would lead us too far to enter more deeply into this reference to early aqueduct physics and into the suggested relation of these physics to the architectural harmonies discussed in the introduction; but one cannot simply say that Vitruvius disregards the aesthetics of the aqueducts.

In Book II he considers the selection and preparation of timber, stone and concrete. He discusses various mixtures of mortar, including those which 'neither the waves nor the force of the water can dissolve'. In Books V and VI he considers the construction of building foundations and frameworks; he points out in detail that in substructures arches should 'discharge the load' of overlying structures and channel this load into piers and thereby into bases.[27] In the aqueduct field these discussions are inherently applicable. The author also expressly says they should be 'committed to memory.'[28] The modern commentary to the aqueduct rules is silent about these discussions and simply asserts that Vitruvius disregards their problems.

In Book VIII on water supply, chapters i-iii contain hydrological observations, chapter iv includes tests for good water, and chapter v begins with a discussion of instruments for levelling an aqueduct grade line. This leads in chapters v and vi to a discussion of aqueduct structures and routings which are mixed with remarks about related topics.[29] The chapter and paragraph numbers are incidentally not authentic and are only retained here to facilitate cross-reference to other discussions.[30] Using the well-known English translation by Morgan, the text reads as follows:

v. 3.   . . . If there is to be a considerable fall, the conducting of the water will be relatively easy. But if the course is broken by depressions, we must have recourse to substructures.

vi. 1.   There are three methods of conducting water, in channels through masonry conduits, or in lead pipes, or in pipes of baked clay. If in conduits, let the masonry be as solid as possible, and let the bed of the channel have a gradient of not less than a quarter of an inch for every hundred feet,[31] and let the masonry structure be arched over, so that the sun may not strike the water at all.[32]

vi. 3.   If, however, there are hills between the city and the source of supply, subterranean channels must be dug ...

vi. 4.   But if the water is to be conducted in lead pipes ...

vi. 5.   ... [and] if there is a regular fall from the source to the city ... with depressions in it, then we must build substrctures to bring it up to the level[33] as in the case of channels and conduits. If the distance round such depressions is not great, the water may be carried round circuitously; but if the valleys are extensive, the course will be directed down their slope.[34]

It is interesting to compare this text with the paraphrase written by the abbreviator Faventinus who found Vitruvius too wordy:[35] the abbreviation omits the entire subject-matter of chapter vi, paragraph 5, while it slightly shortens the other chapters and rearranges their sequence. In effect, it totally alters the meaning of the text. It is also interesting to compare the Vitruvian text with the notes of Callebat.[36] These notes, when dealing with chapter vi, paragraphs 1-5, disregard the Vitruvian discussions on aqueducts and related technology in Books I, II, V, VI, and VIII (chapter v, paragraph 3) and then charge that Vitruvius disregarded important data.

The text and the attempts at restatement and evaluation present undeniable problems. The ancient abbreviator had an impoverished view of what a practitioner needed, while the modern commentator has a more richly developed view of the data essential for the treatise than he himself can keep in mind. We will now take a further look at these data.

Vitruvius and the earlier aqueduct traditions

The earliest elevated aqueducts may be found in Mesopotamia.[37] Groundwater was collected in long, wide infiltration galleries, now known as 'qanaats', and conducted to fields and cities by wide, shallow channels, which at some points were elevated on bridges to carry them over ravines. In northern Iraq a ruin of such a bridge has been found with cuneiform inscriptions dating it to about 700 B.C. It used corbelled arches, which were twenty-five to thirty feet high.[38]

Ancient travellers, such as Herodotus, saw some of the eastern irrigation works,[39] and it is conceivable that the Romans learnt of such bridges, although the surviving contemporary literature does not describe this detail. In any event, however, this was only a small beginning, which was technically obsolete long before Vitruvius, as we will see.

Other types of aqueduct were developed by the Greeks. Vitruvius mentions Herodotus among other writers;[40] for example, he may have known from them or from restatements of their reports about the tunnel constructed by Eupalinus on Samos and its clay-pipe aqueduct.[41] He describes similar aqueducts (VIII, vi, 8-11), and, as already mentioned cites Greek aqueduct 'physics'. The exact content of these hydraulic writings are unknown, since most of those cited by Vitruvius are lost. Most of the actual Greek pipe-line structures are also lost, although some locations are reported.[42] No Greek elevated aqueducts are known.

Nor is there any elevated structure among the many pre-Roman water disposal channels found so far in Italy.[43] However, the Roman constructions of pre-Vitruvian times,[44] provide one major source of information about aqueducts, including elevated aqueducts. We must briefly compare their basic forms with Vitruvius's text and then reconsider the position of this text in the historic evolution of the Roman aqueduct arcades.

The basic construction of the Roman ducts as shown by their ruins confirms the reconstituted Vitruvian text,[45] for example as to aqueduct gradients. As specified in Book VIII, chapter vi, paragraph 1, a very flat, minimum slope was used, which was much flatter than was customary in Greek practice.[46] The ruins similarly confirm the reconstituted text about access shafts of tunnels, which was another detail Roman practice had refined.[47] The practice also confirms the Vitruvian teachings in such areas as mortar cement mixtures.[48] The text specifies the standard dimensions for lead-pipes to be found among Roman remains.[49] In these respects the practice at least reasonably supports the Vitruvian specifications.

According to the artefacts, other portions of the text are inaccurate or indefinite. This applies for example to the uses of masonry in some types of tunnel, the placement of aqueduct reservoirs,[50] and the construction and operation of inverted siphons.[51] Clearly, the text is far from being a perfect description of the practice.

We may also inquire how the Vitruvian text relates to earlier aqueduct writings, where we can find such writings. Aristotle had mentioned aqueduct projects three centuries before Vitruvius, but only to assert that tyrants order such projects to keep their people occupied and poor. When he described his ideal city, he considered wells and cisterns rather than aqueducts.[52] The Ctesibian books cited by Vitruvius are lost and it is only known from Vitruvius and others that they described demonstrations and uses of pneumatic and hydraulic forces.[53] Again, we possess Archimedes's work on floating bodies,[54] but nothing of his on the flow of water. It is conceivable that he wrote a study of the pipe-line aqueducts which existed in this home town of Syracuse when he defended it against the Romans.[55] Vitruvius may have known such a study, since as already mentioned, he writes of both Ctesibius and Archimedes as teachers of aqueduct physics. However, knowledge of elevated aqueducts could hardly come from such a study because the Syracuse pipe-lines only extended through tunnels. Another Greek author mentioned by Vitruvius is Philo of Byzantium, who was a pupil of Ctesibius. According to an eastern tradition Philo's work on pneumatics dealt with the conduction of water; however, the surviving fragments are totally silent about aqueducts, whether elevated or not.[56]

Evidently Vitruvius fully acknowledged earlier aqueduct writings: he mentions more than are known from the surviving literature. If the existing and relatively pertinent remarks of Herodotus, Aristotle, Archimedes and Philo are taken as typical examples, it appears that there had been little of any relevance to aqueducts in general and nothing about elevated aqueducts in the earlier writings. Vitruvius thus appears to have been the first who tried to formulate rules for this field, or at least to outline their essentials.

Vitruvius and the elevated aqueducts of the Romans

The evolution of Roman aqueducts began about 312 B.C. with the first duct, Aqua Appia, which ran entirely underground like the known pre-Roman ducts.[57] The subsequent major ducts gradually developed elevated structures. The layout or course of these major ducts was established about 270 B.C. by Anio Vetus, which extended into the mountainous source area farther east of Rome. The duct described a peculiar S-curve (Figs. 4 and 5).[58] For the first twenty kilometres, it followed the semicircular course of the valley of the Anio River down to Tivoli. Then, substantially maintaining its level, it meandered between the rises and depressions of the Campagna while generally remaining close to the ground surface in a cut-and-cover construction. It emerged only where it spanned small ravines on low bridges. The arches of the bridges used voussoirs with radial joints (as later described in Vitruvius VI. viii, 3); evidently the older corbelled arches were already outdated. Due to its winding and zigzagging or meandering course, the Anio Vetus was much longer than the direct distance from its source to Rome. Therefore it lost much hydraulic head and discharged itself only a few metres above sea-level.[59] Although it was embedded in the ground it must also have collected much heat, since the ground itself becomes hot from the sunshine.[60]

A much more effective design began to be evolved between the middle of the second century B.C. and Vitruvius's time, but the names of the men who conceived and completed it are not recorded. The principles of the new design may be found incorporated for the first time in Aqua Marcia built about 140 B.C. This duct cuts across many detours of Anio Vetus south of Tivoli and further downstream, where it uses tunnels following more direct lines. As a result this duct remains in higher ground, closer to the Alban Mountains. Finally, near Rome Marcia was slightly elevated above the ground on continuous arcades (Figs. 2, 4 and 5). This duct was famous not only for the good taste of the water provided by its spring but also for its coolness when delivered in Rome; Vitruvius specifically mentions the quality of its water.[61]

Thereafter, several smaller aqueducts were built, which either joined Marcia and were superimposed on it, or used underground courses elsewhere (Fig. 4). One of these was called Tepula (lukewarm); another Julia (after Julius Caesar); a third was Virgo which followed a very winding course through deep-cut tunnels near and across the old Appia; and a fourth, Alsietina, brought water from the west. The last three were built in Vitruvius's time, but it is not known whether he took any part in their construction.[62]

The next Roman aqueduct was the major addition of Aqua Claudia, whose construction was completed in A.D. 47 under the Emperor Claudius.[63] Its course was close to the old S-curve of Anio Vetus and Marcia, but its middle section from upstream of Tivoli to Capanelle was significantly different (Figs. 4 and 5). This section began with deep-cut tunnels of several kilometre's length, which extended straight across wide curves of the earlier aqueducts, and continued with surface-cut tunnels closer to the Alban Mountains and correspondingly higher above the River Tiber than the previous ducts. As a result, the downstream section of Claudia from Capanelle to Rome could be and was elevated on high arcades (Figs. 1 and 2). The entire new duct traversed the irregular ground with unprecedented directness. By comparison both Anio Vetus and the famous Marcia have 'timid contours'.[64] In the smoothened S-curve of Claudia, the ancient Roman aqueducts 'reached at last their full development' in the expression coined by a leading archaeologist, that confirmed the opinion of a leading historian.[65] For the first time, spring water from the calcareous Tiburtine Mountains reached all levels of Rome and was delivered in a cool condition.[66]

The downstream arcades of Claudia begin close to those of Marcia where the tunnels of both ducts end. They extend to the entrance of Augustan Rome, the Porta Maggiore, which is now in the middle of the city not far from the main railway terminal. These arcades are about nine kilometres long and generally parallel to one another, but the Claudia still uses a more direct line, while the Marcia crosses it repeatedly at a lower level. Some five years after the Claudia, another major aqueduct, the Anio Novus, was finished. Novus, which began close to the Claudia, followed its bold new course and was carried downstream by the same arcades (Figs. 1, 2, 4 and 5).

The new design, which was built almost straight across the rolling countryside, has been interpreted as expressing an idea of power.[67] If we review its development once more with particular reference to Vitruvius, we can see its beginnings in the Marcia which traversed the irregular ground with a much greater degree of directness. Although this duct became famous, the principle of its construction was disregarded by subsequent builders and not developed any further for more than a century. Then Vitruvius formulated the concept suggested in the Marcia. He wrote that 'the conducting of water' should take place above the plains on raised substructures (VIII. v. 3) and through the hills in tunnels (vi. 3) by means of a channel which slopes positively but minutely, is built to avoid solar heating of the water (vi. 1), and is free of unduly long detours (vi. 5). Within a few decades, the new principle as formulated and developed by Vitruvius reached a state of full development as already described.

Some centuries after Vitruvius, the Aqua Hadriana (Alexandrina) even began to eliminate the basic S-form of the former major ducts and to run east in a substantially straight line, but it was not extended into the mountains and was built on a relatively low level. Some higher aqueducts built in the provinces used long, substantially straight arcades, which adhered to the principle of directness foreshadowed by the Marcia and epitomized by Vitruvius.[68]

Of course, we must not visualize the builders of these ducts as reading Vitruvius's *Architecture* at every moment of their planning and constructing activity. They kept this text in mind, and at hand. As already mentioned, they were provided with an abridgement about A.D. 300 and another abridgement followed about a hundred years later.[69]

The aqueducts were built by contractors and their engineers.[70] These men must have had experience and interest regarding the choice of elevated arcades, surface channels, near-surface or deep-cut tunnels, or possibly inverted siphons. Nevertheless, as mentioned, a considerable degree of uniformity was maintained throughout the imperial period. There had been no such uniformity in the times of the Tepula, Julia, Virgo and Alsietina. We may assume that the change was caused in part by the Vitruvian formulation. It is another question whether the ultimate uniformity was a technical success, or a cause of technical stagnation, or a mixture of both.

According to Frontinus, it was a success.[71] He describes it in words similar to Vitruvius, although he writes of a completed achievement, whereas his predecessor had used the future tense. Vitruvius had summarized some of his aqueduct rules by saying that builders will provide for the conducting of water 'most successfully' when they follow those rules; while Frontinus, referring to the Claudia, called it a work 'most magnificently completed'.[72] Frontinus also noted the 'ancients had brought the water only to rather low levels', which was true of the earliest Roman ducts, especially the Anio Vetus and even of the constructions directly preceding the Claudia. As Frontinus continues, technical progress was achieved by 'subtle exploration of the levelling art' and its use became safe as the empire became free of 'interference by enemies'.[73] This political safety was established by Caesar and Augustus and lasted for several centuries. Vitruvius who lived at the time, was of course not responsible for these political changes and he did not in any way influence 'the levelling art', but, as mentioned, his text did contribute to its new utilization.

The remarks of Frontinus were at one time interpreted as belittling Vitruvius for his supposedly 'ancient' views, particularly on the proper aqueduct gradients. However, these views were presented by confused copyists, not by Vitruvius.[74] Further in his book on *Aqueducts*, Frontius did not charge Vitruvius with such views, since the book is quite in keeping with the proposition that the Vitruvian formulation contributed to the new utilization of improved techniques.

Nevertheless, the imperial aqueducts were not an unqualified success. The masonry channels were made 'as solid as possible', but they could not be made solid enough to operate without continual repairs. The alarming loss of water Frontinus put down to 'theft', but the actual major cause was leakage, which even armies of repairmen could not effectively control.[75] Therefore the Roman aqueducts decayed, largely due to their own shortcomings, although their decay was of course connected in many respects with the general decline of the Roman civilization and empire. After the few centuries of the imperial age, no further aqueducts were built for a millennium.[76]

Vitruvius was still read during and after those times. His reference to Aristotle and others may even have contributed, in some measure, to the gradual renewal of learning. This in turn brought intensified Vitruvian studies during the Italian Renaissance and new attempts to supply water through elevated aqueducts modelled on the Vitruvian principle, particularly near Rome and Paris.[77] Builders then began gradually to recognize that more solid ducts were needed and ducts of welded steel and reinforced concrete as used today developed. As a result, the Vitruvian aqueduct rules have become a matter of historical interest, where study remains difficult.[78]

Notes

The author wishes to thank Professor A. Rupert Hall and Dr. Norman A.F. Smith for their careful review of an earlier draft and for their valued suggestions.

1. Figures 1 and 2 were drawn for this study by Dr. Mark Podwal and were based respectively on a photograph by Fratelli Alinari and on a model in the Museo della civiltà Romana, Rome. Figure 3 has been reproduced by courtesy of the Astor, Lenox and Tilden Foundations at the New York Public Library, Rare Book Division. Figures 4 and 5 were drawn by the author. Figure 4 was developed from C. Singer and others, *A history of technology*, Oxford, 1954-8, vol. 2, p. 669 for general orientation in accordance with Figure 5, which was traced from reduced photographic copies of the large, detailed maps (1:25000) appended to V. Reina and others, *Livellazione degli antichi acquedotti romani*, Rome, 1917. For brilliant illustrations of aqueducts throughout the Roman world, including aerial views, close-ups and all types of architectural drawings, see C. Fernández Casado, *Acueductos en España*, [Madrid], 1972.

2. T. Ashby, *The aqueducts of ancient Rome*, Oxford, 1935. V. Reina and others, *Livellazione degli antichi acquedotti romani*, Rome, 1917, pp. 3f., *passim*. Esther B. Van Deman, *The building of the Roman aqueducts*, Washington, 1934 (reprinted 1973), pp. iiif., *passim*. About the work of Ashby, Reina and Van Deman, see G. Lugli, *La tecnica edilizia romana*, Rome, 1957 (Reprinted London and New York, 1968), vol. 2, pp. 332f. For the continuation of Van Deman's work, see Marion E. Blake, *Roman construction in Italy from Tiberius through the Flavians*, Washington, 1959 and *Roman construction in Italy from Nerva through the Antonines*, Philadelphia, 1973.

3. Vitruvius, *De architectura*, Book VIII, chapter v, paragraph 3 to chapter vi, paragraph 5. (For such citations, this study uses the form Vitruvius viii. v. 3-vi. 5.) The Vitruvian statements in English are quoted from trans. M.H. Morgan, *Vitruvius: the ten books on architecture*, Cambridge, Mass., 1914 (reprinted New York, 1960), mainly pp. 243-5.

4. G. Tabarroni, 'Vitruvio nella storia della scienza e della tecnica', *Atti della Accademia delle Scienze dell' Istituto di Bologna*, classe di scienze morali, memorie, 66 (1976), 18f. Summarizes the aqueduct notes.

5. Cetius Faventinus, *Liber artis architectonicae* in ed. V. Rose, *Vitruvii de architectura*, Leipzig, 1899, p. 285 and on aqueducts, pp. 287-91. A newer edition is H. Plommer, *Vitruvius and later Roman building manuals*, Cambridge, 1973.

6. L. Callebat, *Vitruve, de l'architecture, livre* VIII, Paris, 1973, pp. 144, xlviii.

7. M.H. Morgan, *Addresses and Essays*, New York, 1910, pp. 224-72. P. Thielscher, 'Vitruvius' in Pauly and Wissowa, *Real-Encyclopaedie der classischen Altertumswissenschaft*, 2nd ser., Stuttgart, 1961, vol. 17, col. 458f. Tabarroni, 'Vitruvio', 8-11.

8. Vitruvius v. preface. 1-5. See the editions *M. Vitruvius per Jocundum solito castigatior factus*, Venice, 1511; ed. V. Rose and H. Mullr-Strubng, *Vitruvii de architectura*, Leipzig, 1867; and trans. Morgan, *Vitruvius*. See also Morgan, *Addresses and Essays* and Tabarroni, 'Vitruvio', 11-13.

9. Vitruvius VIII. iii. 27 and vi. 2.

10. Vitruvius II. preface. 4; II. viii. 8-10, 16-18; IV. ii. 2, 3; v. ix. 8; VII. v. 5-8.

11. Van Deman, *Roman aqueducts*, pp. 9-11. Callebat, *Vitruve*, p. x.

12. Vitruvius VIII. iii. 25.

13. R. Bernelle and others, *Académie d'Hippone: comptes-rendus des réunions*, Bône, 1890-1, *Passim*. Also in *Corpus inscriptionum latinarum*, vol. 8, supplement 2, Berlin, 1894, nos. 18858-19110. Also see Pauly and Wissowa, *Real-Encylcopaedie*, vol. 2, col. 307 on Aquae Thibilitanae.

14. Thielscher, 'Vitruvius', col. 419-58. P. Ruffel and J. Soubiran, 'Vitruve ou Mamurra?' *Pallas*, ii (1962), 149f. Tabarroni, 'Vitruvio', 2-5.

15. Vitruvius II. iii. 2; IV. v. 2; VII. vii. 2. Tabarroni, 'Vitruvio', 5f. L. Sontheimer, *Vitruv und seine Zeit*, Tübingen, 1908, pp. 14-16.

16. Bernelle, *Académie d'Hippone*, 1891, pp. xlviif. *Corpus inscriptionum latinarum*, no. 18913.

17. *Corpus inscriptionum latinarum*, no. 19006.

18. Tabarroni, 'Vitruvio', 30f.

19. Pliny the Elder, *Historia naturalis*, book I, comments about book XXXV; book XXXI, xxxi; trans. H. Rackham, *Natural history*, London and Cambridge, Mass., 1947-62, vol. 1, pp. 156f and vol. 8, pp. 412 and 4.

20. Sextus Julius Frontinus, *De aquaeductibus urbis Romae*, trans. C. Herschel, *The two books on the water supply of the city of Rome of Suxtus Julius Frontinus*, Boston, 1899 (2nd ed., London and New York, 1913), p. 18, containing the Latin text of Frontinus's chapter 16 with translation facing it.

21. E. Jüngst and P. Thielscher, 'Vitruv über ... Grundbau', *Mitteilungen des deutschen archaeologischen Instituts, römische Abteilung*, 51 (1936), 144-80.

22. L.B. Alberti, *De re aedificatoria*, Florence, 1485; trans. G. Orlandi, L'architettura, Milan, 1966, pp. 441f., *passim*. G. Cardano, *De subtilitate*, Paris, 1550; ed. Lyons, 1580, pp. 568-71. A. Palladio, *Architettura*, Venice, 1570; facsimile reprint, Milan, 1968, Book 4, p. 4, *passim*.

23. For this trend in art, see F.W. Schlikker, *Hellenistische Vorstellungen von der Schönheit des Bauwerks nach Vitruv*, Berlin, 1940, p. 6 in science, W.H. Stahl, *Roman science*, Madison, 1962, pp. 92-5.

24. V. Rose, *Anecdota graeca et graeco-latina*, Berlin, 1864-70, vol. 1, p. 9. Against Rose and Vitruvius, see E. Oder, 'Ein angebliches Bruchstück Demokrits', *Philologus*, Supplementband, vol. 7, 1899, pp. 343f. and pp. 338, 340, 351. Against Oder on Vitruvius, see Sontheimer, *Vitruv*, pp. 55f, 124 and Schlikker, *Hellenistische Vorstellungen*, p. 7.

25. Vitruvius I. I. i. 7 also reflects more specific aqueduct problems of VIII. vi. 6. It is overlooked by Callebat commenting on Vitruvius's aesthetics and sources in his *Vitruve*, pp. 144, xxvi-xxxix.

26. Alberti (trans. Orlandi), pp. 99, 101.

27. Vitruvius II. iv-vi; v. xii. 2-7; VI. viii. 3-7. Jüngst and Thielscher, 'Vitruv', 156-78.

28. Vitruvius v. preface. 2.

29. Including urban water distribution basins in Vitruvius VIII. vi. 1-2 and lead pipe standards in VIII. vi. 4-5.

30. Ed. and trans. F. Granger, *Vitruvius: on architecture*, London and New York, 1931-4, vol. 1. pp. xxixf.
31. 'Ne minus in centenos pedes sicilico'. The manuscript text 'ne minus ... semipede' was based on confusion of an abbreviation symbol (s or inverted s); see ed. Rose and Müller-Strübing, *Vitruvii*, p. 207.
32. Such arching over is shown by 'c. fornix' in Fig. 3. Sometimes it is replaced by roofing over; see John H. Parker, *The aqueducts of ancient Rome*, Oxford and London, 1876, plate 21.
33. 'ad libramenta', 'to the level [of the substructure]' — not to a level as high as the higher discharge levels in the city. A 'low' level of the substructure for the pipeline siphon is specified in Vitruvius VIII. vi. 5 after the text quoted here.
34. About the latter variant, see N.A.F. Smith, 'Attitudes to Roman engineering and the question of the inverted siphon', in A. Rupert Hall and Norman Smith (eds.), *History of technology: first annual volume, 1976*, London, 1976, pp. 45-71.
35. Cetius Faventinus, *Liber artis architectonicae*.
36. Callebat, *Vitruve*.
37. M.S. Drower, 'Water-Supply', in C. Singer and others, *A history of technology*, Oxford, 1954-8, vol. 1, pp. 531-5.
38. T. Jacobsen and S. Lloyd, *Sennacherib's aqueduct*, Chicago, 1935, pp. 6-18. T. Safar, 'Sennacherib's project for supplying Erbil with water', *Sumer*, 3 (1949), 23-8 and (Arabic, with illustrations) pp. 71-86.
39. I. 185, 193; II. 99; III. 117; ed. F.R.B. Godolphin, *The Greek historians* (including the complete and unabridged historical works of Herodotus translated by G. Rawlinson), New York, 1942, pp. 78f., 82, 128f. 211f.
40. Vitruvius VIII. iii. 27.
41. Herodotus III 60 (ed. Godolphin, p. 190). A. Burns, 'Ancient Greek water supply', *Technology and Culture*, 15 (1974), 389-412 and 'The tunnel of Eupalinus', *Isis*, 62 (1971), 176-85.
42. Smith, 'Attitudes to Roman engineering', p. 52 mentions a number of Greek installations and cites pertinent literature.
43. W.J. Anderson and R.F. Spiers, *The architecture of Greece and Rome*, vol. 2 revised and rewritten by T. Ashby, London, 1927, pp. 30, 123-5.
44. Frontinus (trans. Herschel), pp. 16-22 (chapters 13-20).
45. See note 31.
46. Burns, 'Water supply', 392-4. Reina, *Livellazione*, pp. 31, 75f., *passim*, C. Germain de Montauzan, *Les aqueducs de Lyon*, Paris, 1909, pp. 167-72.
47. Vitruvius VIII. vi. 3 (after the part quoted above, reconstituted in *M. Vitruvius per Jocundum*, folio 81v). About the Greek and Roman traditions, see Burns, 'Water supply', p. 407 and Montauzan, *Les aqueducs*, pp. 285f.
48. Frontinus (trans. Herschel), pp. 155f. Anderson and Spiers, *Architecture of Greece and Rome*, pp. 26f., 32, 41. Jüngst and Thielscher, 'Vitruv'. M.S. Briggs, 'Building-Construction', in Singer, *A history of technology*, vol. 2, pp. 397-418; Lugli, *La tecnica*, *passim*.
49. Callebat, *Vitruve*, pp. 161-3.
50. Callebat, *Vitruve*, pp. 158-60, 170.
51. Smith, 'Attitudes to Roman engineering', pp. 54-9, 68.
52. Aristotle, *Aristotelis opera*, Berlin, 1831-70, pp. 1313 b 20-5 and 1330 b 1-17 (*Politica* v. 11 and VII. 11). Vitruvius refers to Aristotle in other respects in VII. preface. 2 and IX. preface. 2.
53. A.G. Drachmann, *Ktesibios, Philon and Heron*, Copenhagen, 1948, pp. 1-41.
54. Archimedes, *Archimedis opera omnia*, ed. J.L. Heiberg, Leipzig, 1880-1, vol. 2, pp. 345-426.
55. Plutarch, *The lives of the noble Grecians and Romans*, Life of Marcellus, Modern Library ed., New York, n.d., pp. 376-80. Burns, 'Water supply', pp. 389-97.
56. Vitruvius VII. preface. 14. Philo of Byzantium, *Pneumatica*, ed. F.D. Prager, Wiesbaden, 1974, pp. 47-53, 132, 230f.
57. Frontinus (trans. Herschel), pp. 143f.
58. See note 1 for derivation of Figs. 4 and 5.

59. Van Deman, *Roman aqueducts*, pp. 29-66. See also Fig. 4.
60. Figs. 2, 4 and 5.
61. Vitruvius VIII. iii. 1. Frontinus (trans. Herschel), pp. 150-3, 163. Van Deman, *Roman aqueducts*, pp. 67-146. For illustrations, see Lugli, *La tecnica*, plates 51, 1; 64, 4; 171, 4.
62. Frontinus (trans. Herschel), pp. 163-74. Van Deman, *Roman aqueducts*, pp. 23-8, 147-86.
63. Van Deman, *Roman aqueducts*, pp. 187-270. Blake, *Roman construction in Italy from Tiberius through the Flavians*, pp. 26f., 79. Frontinus (trans. Herschel), pp. 175-83.
64. Frontinus (trans. Herschel), p. 154.
65. Van Deman, *Roman aqueducts*, p. 13. Similarly Frontinus (trans. Herschel), p. 176.
66. Frontinus (trans. Herschel), p. 16 (chapter 13); pp. 163, 183 (but note that on p. 165 the names *Claudia* and *Anio Novus* are erroneously printed on a photograph). About the maximum height of Roman aqueducts, see Smith, 'Attitudes to Roman engineering', pp. 65-7.
67. F. Klemm, *A history of western technology*, London, 1959, p. 48.
68. Casado, *Acueductos en España, passim*, mainly on Spanish and African aqueducts. Blake, *Roman construction in Italy from Nerva through the Antonines*, pp. 273-80. Also see K.O. Dalman, *Der Valens-Aquaedukt in Konstantinopel*, Bamberg, pp. 4-18, plates 2-20.
69. Ratilius Taurus Aemilianus Palladius, *De re rustica*; IX. xi; ed. J.G. Schneider, *Scriptorum rei rusticae*, Leipzig, 1795, p. 200. Cetius Faventinus, *Liber artis architectonicae*.
70. Frontinus (trans. Herschel), pp. 177f, 193. Montauzan, *Les aqueducs*, pp. 357-419.
71. Frontinus (trans. Herschel), pp. 186-200; see also note 72 below.
72. Vitruvius VIII. vi. 6. Frontinus (trans. Herschel), p. 18 (chapter 20).
73. 'Veteres humiliori derectura perduxerunt, sive nondum ad suptile explorata arte librandi, seu quia ... infra terram mergebant ne facile ab hostibus interciperentur', Frontinus (trans. Herschel), p. 18.
74. Frontinus (trans. Herschel), p. 194. Ed. Rose and Müller-Strübing, *Vitruvii*, p. 207.
75. Frontinus (trans. Herschel), pp. 256-60.
76. G. Panimolle, *Gli acquedotti di Roma antica*, Rome, 1972, pp. 34f.
77. P. Mastrigli, *Acque, acquedotti e fontane di Roma*, Rome, 1923, pp. 121-67. R.J. Forbes, 'Hydraulic engineering', in Singer, *A history of technology*, vol. 2, p. 691.
78. See notes 31 and 33 above. Difficulties may also be found in the phrase 'eaeque structurae confornicentur ut minime sol aquam tangat' (Vitruvius VIII. vi. 1).
a. The words 'eaeque structurae' may refer back to the several substructures just mentioned in VIII. v. 3 not just to the solid channel structure of VIII. vi. 1. For such referring back one might more strictly expect the term 'illae substructiones; however, the Vitruvian language is flexible in such respects for he also writes 'signa ... inveniuntur' and 'eaeque inventiones' in VIII. i. 3-4.
b. The word 'confornicentur' is capable of different interpretations. In addition to letting the structure be 'arched over', it may mean letting it be 'arched under' by pile-connecting arches. The supporting arches of aqueducts were called 'fornices' as in Frontinus and in an African inscription dated 153 B.C. (*Thesaurus linguae latinae*, VI, 1, Leipzig, 1926, pp. 1125f). According to Cetius Faventinus, *Liber artis architectonicae* (chapter vi), a levelled duct could be provided by tunnels, ground-level channels or by 'structura ... arcuatili'. The latter term, clearly referring to supporting arches, may perhaps be taken as showing the abbreviator's understanding of the Vitruvian 'confornicentur'.
c. The comma before 'ut', inserted in many editions of the Latin text, is unauthentic. (Ed. and trans. Granger), Vitruvius, vol. 1, pp. xxixf.
d. 'Minime' may mean 'minimally' as well as 'not at all'. (Lewis and Short, *Latin dictionary*, under 'parvus-minor-minimus-minime').
e. In view of points a. to d. it is conceivable that the phrase means 'Let these structures be arch-constructed in such a way that the sun can only minimally strike the water'.

f. This in turn could mean 'Let them, for this purpose, be arched over to provide shade for the water, *and* arch-supported as to minimize the length of the duct'.

g. However, the modified translation e. and interpretation f. would not substantially change the Vitruvian rule interpreted by Morgan; it would just provide a specific reason for the subsequent rule in VIII. vi. 5 that circuitous detours should be used only 'if the distance round such depressions is not great'. On the other hand the modified interpretation f. would present a problem, as it would give a double meaning to the word 'confornicentur', which in other Vitruvian Books (for example in v. v. 2) has only one of the two meanings. Therefore the possibility of the modified translation is disregarded in this study.

Two Problems in Fuel Technology

JAMES A. RUFFNER

Since the beginnings of industrialization, modern economies have depended more and more heavily upon fossil fuels — coal, oil, and natural gas. Our nineteenth-century forefathers thought of such materials as sunshine captured in living things, buried and stored up for human use. Since the process takes millions of years, the endowments we have to draw upon are limited and the rate of withdrawal cannot keep increasing. In accordance with the principle of diminishing returns, the extraction rate of any given resource will fall back toward insignificance. Recognition of this prospect of the rise and fall of fossil fuels enables one to understand why finding efficient economically feasible technologies for utilising the continuously available solar energy is of such long term interest.

The Age of Fossil Fuels may be seen as an ephemeral event of less than a thousand years' duration in the history of the human species. The doomsayers note that, at presently compounded growth rates on a globally averaged basis, mankind is only about a century from the peak of consumption. They are pessimistic about the social and military consequences of regions under different political and ideological control facing the hump unequally, with varying endowments under their control and experiencing differing rates of depletion. The technological optimists, on the other hand, note that growth is only a little more than half way to the top and that there is plenty of time and growing incentives to develop adequate substitutes based on the sun (including sunshine, wind, and tides) and on the atom. Some of them are reminded, for example, of the hand-wringing at the middle of the nineteenth century when whales, an important source of illuminating and lubricating materials, were becoming scarce and the Age of Petroleum had yet to dawn. The optimists like to believe that history will repeat itself and the new energy bonanzas will be developed in time to continue growth and increase welfare. The debate is fundamental and could profit by this closer look at the relevant history.

A second major problem in fuel technology is the air pollution that results from uncontrolled emissions, especially of the products of incomplete combustion, from large or densely packed sources. Each fuel creates its own problems of combustion and emission control. The ones associated with coal have been a long standing bane of civilized people. The high volatile, high sulphur, high ash bituminous or subbituminous 'soft' coals readily available to the heavily industrialized regions of Europe and North America which constitute a large portion of our fossil energy reserves have posed unusually serious problems.

Various furnace management schemes involving the proper admission and mixing of air and either mechanical or more skilful hand-stoking were devised from the time of James Watt. Early smoke-abatement programmes were educational campaigns based on the diffusion and demonstration of these techniques, backed by the common law against nuisances or specially enacted police powers. The campaigns achieved only temporary success; in part because of the rapid increase of industrialization and the failure to make significant headway against domestic smoke. Greater ease of application and effectiveness of control were promised than could be achieved. The legislative and enforcement process left much to be desired and the momemtum of these campaigns tended to be lost.

Although smoke washing and scrubbing techniques date back one hundred and fifty years, the problems with acid corrosion were difficult to overcome. Only now in their fifth or sixth generation do scrubbers seem likely to achieve their early promise, even if at a rather high price, with the economic — and energy — trade-offs still very much at issue. The dawning of the Age of Electricity nearly a century ago made possible two new strategies for cleaner air: electrostatic precipitation of smoke and small particles and the general replacement of a large number of small inefficient fires for heating and power purposes by a few large, well-managed fires in central power stations to provide steam heat and electrical power. This replacement may seem strange in the light of present day problems with pollution from central power stations, but it made good sense in the context of sources existing then, but now largely eliminated by the success of electrification and the substitution of cleaner burning fuels — fuels which today show characteristics of increasing scarcity.

An episode of particular significance occurred early in this century when various schemes were being advanced to force electrification of locomotives on the complex Chicago railroad terminal system. The effort in Chicago was part and parcel of the Progressive movement, even though it provoked sharp debate on what constitutes progress. However, before considering the confrontation in Chicago of sixty years ago which provides one comparative perspective on the problems of energy technology in modern society and which forms Part II of this paper, it has seemed appropriate to examine in Part I a different comparative perspective, where the issue is one of probable shortage. We are but too familiar with the notion of a world made cold and immobile by a shortage of fossil oil; the mid-nineteenth century faced (as it seemed) a future that would be dark, or at least only candle-lit, because of a shortage of animal illuminant oil. The arguments about this situation, and their outcome, are, like the Chicago railway question, full of relevance for our own day.

Part I. The whale-oil crisis re-considered

In 1857, only two years before the start of development of the petroleum district of western Pennsylvania, *The Scientific American* noted with alarm:

> The whale oils which hitherto have been much relied on in this country to furnish light, are yearly become more scarce, and may in time almost entirely fail, while the rapid increase of machinery demands a large portion of the purest of these oils for lubricating.[2]

The story of this crisis has recently been polished into a neat, albeit simplified, object lesson for doubters of the cornucopian theory of resource economics and the power of technology induced by free competitive markets to solve any resource problem. The technological point of view was presented by Gould, Inc. in recent full-page advertisements and reprinted in a widely-distributed pamphlet. The Gould version of 'Whale Oil, Arab Oil, and No Oil' begins:

> No one who lived through it can forget the fuel shortage of 1974. But few remember that today's energy crisis is not the first one in our history.
> Up until the middle of the last century the lights of the world were lit with whale oil. Machines were lubricated with whale oil.
> Suddenly there wasn't enough. Whales were being killed faster than they could spawn. An energy crisis was imminent. Fears that the worst could happen were commonplace.
> Then an unexpected thing happened. Technology came to the rescue. As whale oil supplies dwindled, the growing petroleum industry took up the slack.
> Not only did the petroleum solve the problem — it was *better* than its predecessor.[3]

In earlier versions, promulgated by Dr. W. Philip Gramm, a professor of economics at Texas A & M University, and repeated by Robert Lund, General Manager of Chevrolet, the emphasis was on the market inducements to technological innovation.[4] 'Gramm claims that:

> The whale oil crisis is a case study of how the free market system solves a scarcity problem. The end product of this process of discovery and innovation is the Petroleum Age is which we live. We owe the benefits and comforts of the present era to free enterprise and the scarcity of whales.[5]

Long lead times, hard work, an unexploited resource base, and good luck are also required to maintain the flow of energy and other materials in a growing economy. According to this theory, nature abounds with unexploited stores that human ingenuity can sooner or later command at a price. The long lead times for research, development, and diffusion and the associated hard work are triggered off by signals in the free price mechanism. Then again there are elements of serendipity and of luck in getting everything together in timely fashion to prevent periods of social and economic disruption. According to the modern economic, or cornucopian, view of resources implicit in these accounts:

> The real question . . . is not whether resources exist but at what rate different sources of supply will become available to man in the sense of becoming economically feasible to recover. Natural materials do not become resources until they are combined with man's ingenuity. Over time the record is impressive. Mineral resources have become more and more widely available despite (and partly because of) growing rate of consumption.[6]

In the USA the resources for illuminants and lubricants broadened considerably in the half century before the Civil War as the markets of a growing and increasingly literate and prosperous population demanded light and machines.[7] Oil was squeezed or refined from whales, seals, fish, animal fat, seeds, wood, coal, and ultimately petroleum. Gas was also made from many of these resources. Candles were made from tallow, wax, and other animal by-products including, pre-eminently in quality, head matter or spermaceti of sperm whales. At the other end of the economic spectrum, pine cones and rags saturated in unrefined fat were useful, and many a person made do with the light of the fireplace or with no artificial light at all. The standards of excellence and safety down to the Civil War, however, were set by the whale-based products. Still, it is not entirely accurate that 'Since there were no good substitutes for these oils as sources of light, the world's supply of artificial light depended almost exclusively on the whaling industry'.[8]

The supplies of the major oils and coal-gas during the period and up to 1880 can frequently be little more than guessed at because of the fragmentary records. Tentative values are provided in Figure 1. Note that production is plotted on a logarithmic scale where each major division is ten times the lower one.

The period of the 1820's through the 1840's was known as the Golden Age of Whaling.[9] The American industry became increasingly concentrated in the New Bedford, Massachusetts area and its range of exploitation was extended to all parts of the Pacific and Indian Oceans. In 1848 the whalers tapped the contemporary equivalent of Alaskan North Slope oil fields as the Arctic Fishery above the Bering Strait was opened. The length of the voyages increased from two or two and a half years to four years or more. The sizes of ships, crews, and other investments grew proportionately. With greater exposure to storms and growing desertion and mutiny, investments in whaling were put at increasing risk and the returns were uncertain and long-delayed. (The California gold rush of 1849 was a disastrous invitation to desertion or

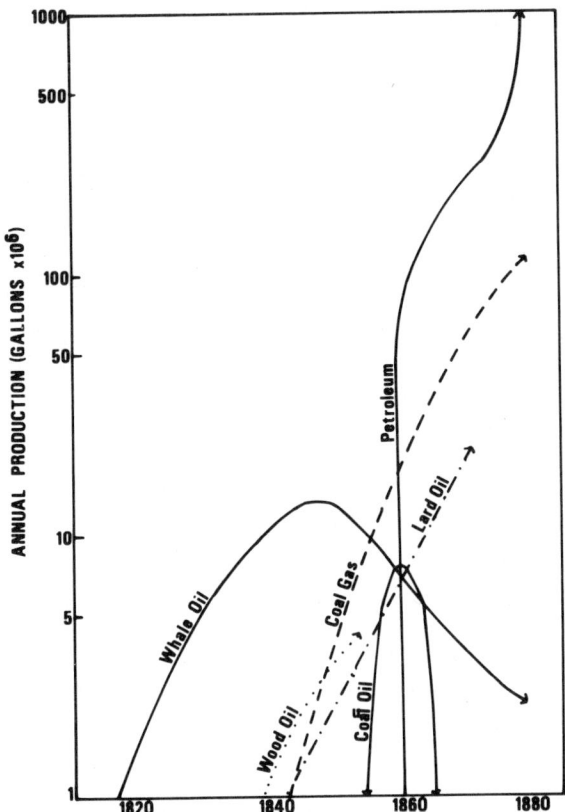

Figure 1. Production of oils and gas in the USA, 1820-1880 (Author's estimates).

worse.) The industry declined because of the increasing scarcity of both whales and capital. Many investors turned to the safer and more easily supervised cotton textile industry that was coming to rapid maturity in New Bedford, the Fall River district, and other nearby southern New England locations.

Exploitation was directed mainly at two species, sperm whales and right whales. 'Right' whales surfaced after an hour or so's struggle, the 'wrong' whales were larger and sank when dead so that they could not be handled from the small dories sent out from the mother ship for the hand-harpoon kill. Thus, the vast resources of the larger species such as whale-bone, blue, fin, and humpback were impregnable until after 1860 when harpoon cannons, steam winches, and air pumps to restore flotation were introduced aboard the main whaling ships. These innovations however were Norwegian and leadership passed into their hands as the American industry continued to decline. The Norwegian and British whaling industries did not yet have to contend with North Sea gas and oil.

During the Golden Age as well as the crisis period, wholesale prices for whale oil fluctuated by as much as fifty per cent from year to year, with the annual increase or decrease averaging about twelve per cent. Corrected for fluctuations in the cost of living, imperfect as such an index for the period must be, these price changes followed amazingly regular cycles of five to seven years.[10] The long term trends from cycle to cycle show that, in real terms, the wholesale price trend of ordinary whale-oil was almost steady only falling slightly through the 1830's, while the trend curve of the premium sperm-oil increased by one or two per cent a year (see Figure 2). In the 1840's, the trend in real prices was towards an increase of two or three per cent a year for both classes of oil. In any of these cases, the noise of the year to year fluctuations must have obscured the message to all but the most perceptive observer until the 'crisis' was nearly at hand. By the late 1840's, however, the trends in prices were soaring by ten to fifteen per cent a year, so that the message became less difficult to read.

Production increased by ten per cent a year, peaking at about 5 million gallons per year around 1840 for sperm oil and at about 14 million gallons per year between 1845 and 1847 for the total of all classes of whale-oil. Due to falling exports, the net amount retained for domestic consumption or storage continued to rise erratically to about 10 million gallons in 1853 and 1854. Total annual production was then down to about 12 million gallons. From 1854 the industry declined steadily to reach insignificance by the turn of the century. After the collapse in prices following the Civil War, caused by effective competition from petroleum-based substitutes, the oil commanded a price that barely covered the handling cost and whales were hunted mainly for bone for umbrellas, corsets, and other fineries. Just after the turn of the century one historian of the industry, noted that 'few industries offer an opportunity for such a complete study of the rise and fall.'[11]

During the rise of the industry, the United States population increased by three per cent a year and, perhaps more to the point, the population of urban places with more than 2,500 inhabitants increased by six per cent a year.[12] Even measured against this urban increase where the markets were best organized and the affluent middle classes were concentrated, whale-oil increased its penetration of the market substantially despite the rise in prices. The New Bedford whalers were certainly bringing in the 'good stuff'. Yet, the peak supply was large enough to fuel only about 6-700,000 domestic lamps producing a full seven or eight candlepower for 1,000 hours a year, or perhaps one million lamps at reduced power.[13] Choosing this high number, and making no allowance for commercial and street lighting or lubricating and other non-illuminating purposes, since there were four million households in the USA, many using more than one lamp, it is unlikely that more

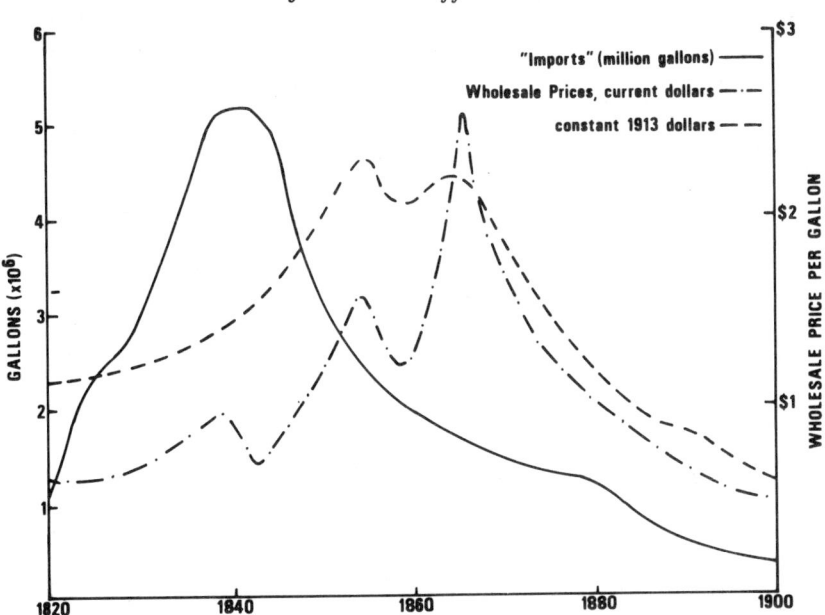

Figure 2. Sperm oil price trends in the USA, 1820-1900.

than fifteen per cent of American families commonly burned whale-oil in the years of peak supplies. Spermaceti candles might have graced 250,000 households at their peak, down to perhaps half that number in the early 1850's.[14] The market for whale products was restricted to middle and upper class households and businesses.

Many other materials were developed as light sources before the Civil War to facilitate the exploitation of the technological capabilities of vast resources of coal, wood, seed crops, and waste animal fat. Some of these products such as coal-gas and colza-oil from the seeds of rape-plants offered high quality light at a high price in direct competition to the prestige whale-oil market. Other products opened new markets. All benefitted from the rapid population growth of the nation and the rising demand for artificial lighting as literacy spread. High prices *per se* (intrinsically an indication of scarcity) rather than the rising prices of whale-based products may have provided initial incentive enough to develop efficient refining techniques for other more abundant raw materials. Scientific curiosity and the Biblical desire to master Earth also played a rôle in preparing for the eventual changeover long before the crisis hit.

Gas manufactured from coal was introduced in Baltimore in 1818, following successful British innovations in what was one of the first truly science-based industries rather than an extension of earlier craft-based trades. Only seven gas light companies were formed in the USA before 1840, by which date various technical problems in the purification of gas (based on theoretical chemistry) and its effective distribution and metering had been overcome by the British and copied in the States. Subsequently, the gas industry expanded rapidly to supply street lights and to serve a few middle class urban homes that might otherwise have afforded whale-oil. The burgeoning industry provided the model for Edison's electric lighting system powered by centralized coal-fired steam engines or turbines which eventually displaced the coal-gas system for lights.

The few scattered figures available for gas production in the 1860's and again after 1890 reveal that the long-term growth rate was seven per cent a year, with product doubling every ten years. More to the point in estimating the number of households served for the period up to 1880, when the cooking and space-heating demand began to provide much of the growth and further technological changes altered the industry, would be the numbers of companies and the population of the cities served by them. The census of manufacturing for 1860 lists 221 companies, while an industry journal survey for 1859 sets the number at 297 companies.[15] Part of the discrepancy may be explained by the existence of many gas-producers with industrial plants that served nearby buildings with excess gas without technically being in the gas business. The industry survey indicated that the 297 companies served 227,665 private customers in population areas of 4,857,000 persons or approximately 900,000 households. Thus, the average company had less than 800 meters, many of them serving business and institutional customers, and the number of households reached in the service area would certainly have been less than twenty-five per cent. Philadelphia, with city-owned plants, may have had the highest degree of domestic service. At the end of the present period of survey in 1880, Philadelphia contained 146,000 dwellings and perhaps 18,000 commercial, industrial, and institutional buildings.[16] The city-owned plants had about 99,000 customers, of whom no more than about 80,000 would have been domestic, although the market penetration could have been well over fifty per cent.[17] Less certain figures for Cincinnati in 1880 suggest that there were 10,000 or fewer domestic customers out of a total of 14,000 customers serving a population of some 50,000 households, or perhaps a twenty per cent market penetration in one of the oldest and wealthiest midwestern cities.[18] No overall figures are available for 1880, when no census of manufactures was taken, but simple-minded interpolation between available scattered figures before and after indicates the existence of around 550 gas plants averaging fewer than 1,000 domestic customers each. It is not likely that more than half a million households in the whole USA had gas lights as late as 1880. This figure would equal one-third of the households in the principal cities served, or fewer if all cities were known and totalled.[19] It represents five per cent of all households nationally. The prices and the nature of the necessary service areas were such that 'the people' were not yet served by gas companies even though the industry grew at an enormous pace. The diffusion of cheaper water-gas technology (an innovation of the mid 1870's) and the introduction of more efficient burners, especially the gas mantle after 1890, coupled with the stimulation of competition from Edison electric companies, may have changed the market penetration somewhat. The social history of our fabled gas-lighted cities is yet to be written.

In the 1830's, mixtures of wood-alcohol and turpentine, such as Camphene, were marketed as cheap, bright-burning, but explosive lamp oils.[20] Despite the hazards and a high rate of consumption, such burning fluids were said to be extremely popular until finally displaced by cheap kerosene. (Unfortunately, early kerosene was frequently adulterated with gasoline, a by-product waste of its manufacture, to produce an equally dangerous mixture.) About 1840, hog-packers in Cincinnati, the bacon and lard capital of the world, adopted French scientifc discoveries of the previous decade to upgrade certain by-products to make improved lard-oil and sterine (adamantine or star) candles. These high-quality products were marketed at prices between the dangerous burning fluids or low-quality tallow candles and the expensive whale-oil and spermaceti candles. Lard-oil continued to be manufactured throughout the century, its product value increasing sharply after 1860, but its fate as an illuminator is obscure.[21] Presumably it faded as kerosene became plentiful and cheap. At least for a while, it successfully challenged sperm-oil as a lubricant.

At various times, *The Scientific American,* in its typically overstated fashion, noted that each of the new products based upon coal, wood, or hogs was in its turn 'rapidly superseding whale oil.'[22] Price softness in whale-oil in the early 1840's may partially reflect this competition, but a major economic depression offers adquate explanation. It seems fairly likely that for a time many families enjoyed improved standards of lighting or cheaper substitutes for the whale-based products. Yet, based on very sketchy information, it is possible that the supply of all of these new products increased sufficiently through the 1840's to reach perhaps only a quarter of a million households, not enough to keep pace with population growth, let alone displace the 'good stuff' for long, even if that were the dream of the entrepeneurs.[23] Coal-gas provides the exception. There is no reason to believe that the situation, relative to population size, had changed much by 1860.

The natural theology of the day reassured many developers that there were keys to open Earth for the riches providentially stored in it for human use. As early as 1846 Dr. Abraham Gesner, a Canadian physician, began to experiment with the distillation of Trinidad asphalt to obtain a lamp oil. Later, as indicated below, he was instrumental in opening up the shale-oil industry in this country. As for his natural theology of asphalt, which he regarded as 'altogether inexhaustible', he noted with supreme satisfaction, 'for what purpose nature had formed such vast quantities of bituminous matter, which still continues to flow from the earth, was a problem not readily solved, until this discovery [of distillation and refining], which brings [it] into operation for illuminating purposes, to which it is admirably adapted.'[24] Gesner, of course, was far from alone in his view that resources were more the work of God than of the economic system which at best was His handmaiden. Thus, one writer observed, 'nature does not display all her treasures at once, but opens one storehouse after another as man's needs may require.'[25] Jumping ahead of the story to the development of petroleum, another writer observed, 'in compensation for privation and poverty [in western Pennsylvania], our Kind Father in Heaven has caused the rock to pour us out rivers of oil and thus given us at once magnificent light for our dwellings and a source of honest wealth.'[26] But clearest of all, *The Scientific American* asserted:

> It is very evident that the earth was prepared with the special end in view of being man's abode, and the Great Architect of it has laid up stores in the bowels of the earth, from which man is to be supplied with light and heat, when our forests shall fail, and the whale cease to be chased by the daring mariners of Nantucket.[27]

In retrospect, the key to the storehouse and the most significant competitor with all other illuminants and lubricants came from the innovations that led to the short lived coal-oil industry of the 1850's and early 1860's.[28] A. F. Selligue began distilling oil shales in 1834 for a product to enhance the heating and illuminating properties of manufactured gas. (Later naphtha and, not very successfully, gasoline were used for the same purposes.) In 1838, Selligue patented his product as a lamp oil, and within a few years he was operating three refineries in his native France to supply oil both for gas manufacture and for lamps. In Great Britain, in 1847, James Young began to exploit an oil seepage for lubrication purposes. To achieve a larger and more reliable resource base, he turned to low temperature distillation of Scottish Boghead cannel coal, a highly volatile coal — sometimes considered a shale — used extensively in gas manufacture. By 1852, he had obtained US patents for the distillation and refining of bituminous sources that were broad enough to control all coal-oil processes in the country. By 1849, Abraham Gesner had turned from Trinidad asphalt to develop the commercial possibilities of New Brunswick albertite, a material that can be termed loosely a high-grade oil shale. He developed techniques to produce both a high quality manufactured gas and a liquid with qualities similar to Young's paraffin oil. He set up business in New York City in 1854, selling his main product under the trade name of Kerosene. Other pioneers struggled to develop coal-tar based lubricants, receiving support from a Boston producer of sperm-oil and candles who felt the need to diversify as a hedge against further price rises in whale products. In 1856, this company and others followed Gesner's lead (who meanwhile had lost a patent battle with Young) into the coal-oil business.

Following scattered but successful efforts at development over the previous twenty years in Europe and the USA a major industry was created with incredible speed. After commercial operation started in 1854, industrial production of Kerosene and similar brand-named products reached about eight million gallons in 1859-60 or enough to keep some two million lamps burning (at less than full power) several hours a day for a year in a more or less million households.[29] A coal-oil mania as well as a business panic was sweeping the country and, at the beginning of the Petroleum Age, the industry suffered from overcapacity. Price softness was evident in the whale-oil market as coal- or shale-based oil flooded the depressed market, bankrupting several producers including Gesner's former pioneer company and driving down their own prices. And yet, in 1860, there were more than five million households in the country. Only time would tell whether coal-based kerosene would turn the trick after the economy recovered and the industry could get going again.

The cost of lamp-light equalling eight candlepower with coal-oil or 11 candlepower with wood-oil cost about 2¢ per hour at 1859 prices and 1¢ or less per hour at 1860 prices. Similar illuminating power from coal-gas cost 1.5¢ to 2.5¢ per hour, from sperm-oil 5¢ per hour and from high-quality candles 5¢ to 10¢ per hour. As typically burned in pairs, the best candles cost a penny or two an hour. The sub-candlepower light of one ordinary tallow candle cost about o,3¢ per hour.[30] The relative cost of such illumination can be measured against typical urban working wages of $300 to $600 per year, half of which went for food.[31] Poor people always practiced perfect economy of whale-based products and their high or middle priced substitutes. Home made tallow candles, pine cones, and the ill burning home-made products of unrefined animal fat or seed-oil were the most common light sources of all, until kerosene became really cheap.

The origins and rise of the petroleum industry are well documented. It is frequently claimed that we owe the Petroleum Age to the scarcity of whales. A full account, however, must consider the role of other industries and the serendipitous nature of many of the events leading to the drilling of the first oil-wells. Petroleum or rock-oil was noted initially for its propensity to spoil salt wells and touted for its remarkable curative powers. Early attempts to use it as the basis for lubricants and lamp-oils were abandoned in favor of coal-and shale-oil, evidently to exploit a more secure resource base. The joint-stock company that eventually led to the first well deliberately drilled for oil was founded originally to exploit its value as a medicine. Only after an investor group insisted upon a scientific analysis did Professor Benjamin Silliman, Jr, establish its potential value as an illuminant. The company still had no clear idea of how to proceed technically to recover oil and ignored completely the problems of refining it. Delay and uncertainty in the venture intimidated the investors and the business panic of 1857 nearly finished the effort. In 1858, Edwin Drake, an unemployed clerk and railroad conductor, and ersatz colonel, was

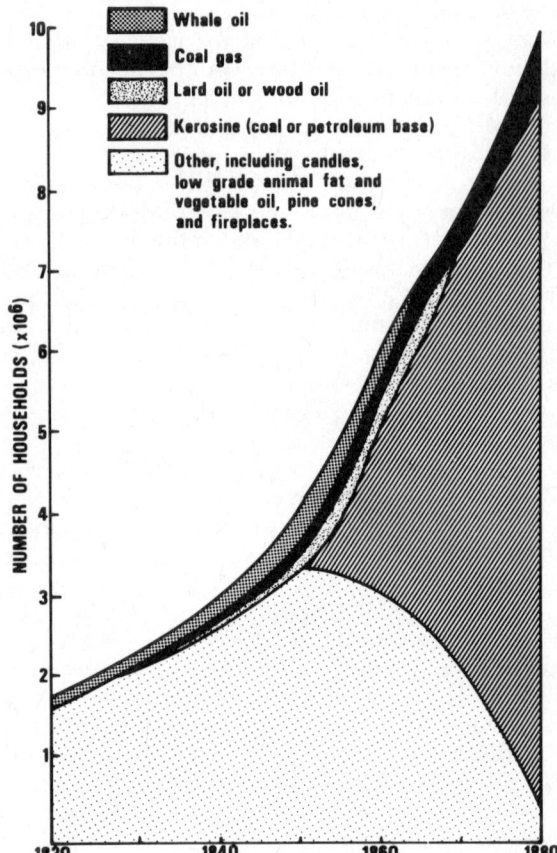

Figure 3. Prime household illuminants in the USA, 1820-1880 (Author's estimates).

hired to get the project going again. Conventional trenching and well digging techniques resulted in some oil and a lot of mud. Since oil wells had been drilled inadvertently in the search for more valuable brine, a succession of salt well drillers was unsuccessfully employed in desperation. The whiskey consumption was higher than the oil production. More mud holes resulted and, in a day of imperfect communication, the order to shut down and cut losses was sent, only to arrive a day or two after the dogged persistence of the new and more sober driller, W.A. 'Uncle Billy' Smith, had snatched victory from defeat. As one oil historian has noted, 'to the very end the project to drill for oil was kept alive by a lengthy chain of extraordinary circumstances'.[32]

The 'Drake' well had many imitators. How soon other speculators would have ventured if the project had failed cannot be known. Perhaps the dawn of the Petroleum Age would have been delayed only a few months or years. Yet, other abundant sources of hydrocarbons were undergoing rapid development for uses in solid, liquid, and gaseous form. As it was, after many technical problems of refining had already been solved, petroleum substitution in the coal-oil refineries began in late 1860 and was widespread by 1863. Many refineries were founded to handle petroleum exclusively. Coal- and shale-based oil had nearly vanished by 1865 when for all practical purposes in the USA it was shelved until next oil crisis.[33] Supplies of crude oil reached the 8 to 20 million gallon range during the first full year and perhaps 100 million gallons in the second year. About half of the crude oil may have been distilled and refined as kerosene. Typical household consumption was in the order of ten gallons per year, at a price nearly everyone could afford.

Estimated trend curves for the production of the various fuels in equivalent units are depicted in Figure 1 above. My crude estimates of the impact of these materials on American households are summarized in Figure 3. The first truly revolutionary impact was that of kerosene. In the diagram, the impact of kerosene acquires something of the shape of an inserted cornucopia a fitting symbol of the new age. The disposal of all the by-product gasoline made in the process is the next problem. But that is another story.

Part II. Pollution and progress

The great World Expositions between 1893 and 1904 held in Chicago, Paris, and St. Louis helped to make commonplace such delights of the modern age as Ferris wheels and other Midway attractions, hot dogs, hamburgers, ice cream cones, and iced tea.[34] They also gave our forebears dazzling displays of electric lights and awesome demonstrations of electrical power, thereby helping to quicken the diffusion of the relevant technologies which were increasingly seen as symbols of the modern age. Farm folk such as the novelist Hamlin Garland's parents were overwhelmed by these displays, while intellectuals were lifted to higher planes of fact and fancy, as Henry Adams in his essay, 'The Dynamo and the Virgin'.[35]

The progressive era was at hand and conservation and reform leaders were intensely optimistic as the new realms of science and technology appeared to open up unlimited opportunities for human achievement.[36] The concept of planned and efficient progress lay at the heart of both the conservation idea and the City Beautiful movement. The coming electrical age would help to realize both of these ideals. This belief was promoted vigorously in Chicago and the American 'Midlands'. Eminently practical businessmen and engineers became hard-sell advocates of the economic, social, environmental, and resource-conserving benefits of electrification.[37] High efficiency, fuel saving, smokeless central power stations would be interconnected in load-sharing grids to serve diversified demand, including central-heating from the waste steam. Diversified demand would yield an economically favorable high average load factor and replace the much less efficient, more costly and polluting individual power plant. William F. M. Goss, Dean of Engineering, Purdue University, set forth in 1906 many of these ideals in this provocative view of the modern smokeless city:

> A city, to be made smokeless ..., would first seek to fix limits defining the area to be controlled. Within this area would be developed a series of power and heating plants which would be spaced upon a system of squares in the business portions, at intervals of a mile or a mile and a half, and in the residential portion at intervals of two miles. From these several stations would go out currents of electricity for all power and light needed by the city. From certain of them steam at high pressure for industrial purposes would be distributed over the limited areas and from all of them would go out steam or hot water for heating. ... Because of their size and perfection of equipment, all would be operated by smokeless fires. All small fires, which at the present time serve for heating and power in individual buildings, would cease to exist, and large fires under boilers of great industries and in furnaces of metallurgical establishments, would be made smokeless by means which would enhance their economy of operation. Railroad trains passing through the controlled area would be drawn by smokeless locomotives, and above and around the city a clear atmosphere would contribute to the cleanliness of all things and to the comfort and peace of mind of all its people.[38]

Goss, who was a world authority on steam locomotive economy, had various design and operational changes in mind that would greatly reduce the smoke from steam locomotives, but others were sure that the future belonged to electric locomotives. The superiority of central electric tramcar and light-rail systems over other forms of traction was fairly well established by the time of the St. Louis Fair in 1904, although gasoline, gasoline-electric railcars or 'doodlebugs', the forerunner of the diesel-electric locomotive, were under development for use in low density service areas. The adaption of central electric traction to heavy rail service seemed inevitable to many electrical advocates. Thus, speaking softly in 1904, while electrification was underway in New York City, one Chicago civil engineer told a local audience:

> How rapidly this change from steam to electric propulsion of trains will be carried out by the trunk lines is hard to predict, but that the change will be made seems ... to be a foregone conclusion. The high speed electric railway is with us to stay. How great the change it will bring about in the civilization of this country is not possible to predict.[39]

Later that year, Bion J. Arnold, Chicago's leading specialist in electrical traction, called attention to the big stick:

> That electricity generally will be used on our main railway terminals, and ultimately on our main through lines for passenger and freight service, I am convinced, but I do not anticipate that it will always be adapted on the grounds of economy of operation, neither do I anticipate that it will come rapidly, or through voluntary acts of the owners of steam railroads, except in special instances.

He continued:

> In view of the present state of the art of electric railroading, the right to vitiate the air of our congested districts by the emission of large quantities of carbon dioxide from the stack of numerous locomotives is as questionable as the right to foul air by the stenches from our stockyard.[40]

The concern for the health effects of carbon dioxide at concentrations only two or three times natural background levels reflects a fading tradition in medical thought about vitiated air as the germ theory gained wider acceptance, but concern for other air pollutants was rising. Smoke, soot, fly ash and other forms of grit, sulphur dioxide and acid rains, coupled with the concern for fuel economy, were leading to a renewal of the clamour for cleaner air in all parts of the industrialized world, at a time when the ambient measures of those old standbyes — particulates and sulphur dioxide — were an order of magnitude higher than comparable measures today.[41]

In the period just before World War I, the newspapers of America carried numerous stories and editorials on the subject and many writers, echoing themes like those expressed by Goss, believed that clean air was just around the corner.[42] Many civic patriots, however, regarded such a hope as superfluous. Thus, Edmund James, president of Northwestern University, detected the 'strange beauty and sublimity ... which lurks everywhere under the smoke and dirt of our city life, a life which seems to the darkened vision of a supercilious aesthete only sordid and mean'.[43] He also noticed 'how little after all, man has defaced the natural beauty of lake and sky and river and woods' and how some of the 'ugliest creations seem to increase or heighten their very beauty'. Many other people were proud of their city despite the smoke, or perhaps even because of it. The popular song, 'The March of Chicago', offers the refrain:

> Three cheers for the mighty city,
>   As she marches on her way,
> With her banners high in the smoke-filled sky
>   And her face turned toward the day;
> Marching along, two million strong,
>   Three times three cheers for Chica-go![44]

Chicago, the second city in America and soon to be, according to the school-book version of the Chicago Plan, the first city of the world — Chicago, turn-of-the-century shock city and heir-apparent to Manchester, England, as 'The New Athens', the symbol of the age — Chicago, the Great Central Market — Chica-go, the 'I will' city![45] Yet, how starkly the grim reality of the host 'Gray City' with unplanned growth at its core stood in contrast with the dignity, beauty, and convenience of the transitory 'White City' of the 1893 World's Fair with consummate planning and science and technology at its core.

Early in this century, a major collision of interests took place in Chicago. The urban realists saw life and economic progress. The urban reformers believed it was not their vision that had darkened, but to use the novelist Booth Tarkington's apt phrase, it was their town which had in less than a generation 'darkened into a city'.[46] Chicago was noted for its dirty water supply, its dirty air, its stinking stockyards, its frequently chaotic railroad terminals, its incredible street congestion, and the dismal character of much of its architecture to say nothing of its rotten politics. Reform was required on all fronts and just after the turn of the century the prospects looked good. Progressive politicians were coming to the forefront, and the famous 'I will' spirit was being uncorked, even as civic patriots glossed over the problems. The Chicago Sanitary and Ship Canal had been completed in 1901 at a cost of $60 million. The annual level of spending on new buildings had approached $100 million by 1910. The various metropolitan park boards were spending a total of $6 million per year in 1908 and the system was growing. A mass rapid-transportation plan, including an underground electrical freight system within the Loop district, was being implemented with projected costs of $275 million over the first quarter century. Beginning in 1899, legislation was passed to force the trunk-line railroads and switching companies to elevate or depress portions of their track to eliminate crossing jams and accidents at a projected cost of $150 million spread over two or three decades, and another $150 million would be required for the improvement of terminal facilities.[47] Indeed, there were at least a half-dozen schemes to rationalize the chaotic terminal system.

The Chicago Plan was unfolding. Smoke would yield to the forces of progress as well. Tarkington was not so sanguine. At the height of the ensuing controversy over smoke abatement, the Gentleman from Indiana was at work on his first attempt to write the urban novel:

> There is a midland city in the heart of fair, open country, a dirty and wonderful city nesting dingily in the fog of its own smoke. The stranger must feel the dirt before he feels the wonder, for the dirt will hit him instantly. ... At a breeze he must smother in whirlpools of dust, and if he should decline at any time to inhale the smoke he has the meager alternative of suicide.
>
> The smoke is like the bad breath of a giant panting for more and more riches. ... 'I will make Wealth! I will sell Wealth for more

Wealth! My house shall be dirty, my garment shall be dirty, and I will foul my neighbor so that he cannot be clean — but I will get Wealth!' And yet it is not wealth that he is so greedy for: what the giant really wants is hasty riches. To get these he squanders wealth upon the four winds, for wealth is in the smoke.[48]

Chicago was the midcontinental railroad centre. The celebrated 'closing' of the American frontier — to which the historian Frederick Jackson Turner drew national attention — merely served to intensify the development of the interior of the nation as did the irrigation systems of the arid west, the spread of the wheat belt, the scientific intensification of cornbelt culture through the application of genetics, and the development of the co-operative agricultural extension network and other forms of technological management. Chicago and its rail traffic prospered. The Panama Canal was a factor in the development of the Mississippi Valley and provided a second-order stimulus to the growth of Chicago and its railroads, even as it weakened them to the west. The opening of American trade to the Far East was also seen as a factor in the city's growth.[49] Chicago was truly the great central market. It was the crossroads of the nation and much of the trading world. As one country editor wrote in 1897 from his more limited perspective:

> Land O Goshen; it's a Horn of Plenty. How would the nation clothe itself, feed itself, keep warm if it were not for the railroads of Chicago.[50]

The development of Chicago's railroad terminal system, like that of the city at large, had been achieved in haphazard piecemeal fashion. The result was appalling inefficiency. Terminal congestion caused delay in shipping. Most of the passenger facilities were inadequate and outmoded. Much real property was tied up in unproductive uses. A 'Chinese Wall' of railroad facilities blocked the establishment of through streets and the expansion of the central business area to the south of the Loop. Numerous civic and railroad leaders pressed for reorganization of the rail system. Thus, in the winter of 1906-1907, during a particularly serious failure of the railroad system all over the country, Frederick Delano, known both as an architect and planner and as president of the Wabash Railroad, was joined by no less a person than James J. (Jim) Hill, the empire builder, in stressing the paramount need for improved railroad terminals, rather than simply acquiring more cars.[51] Such pleas were crystallized in the Chicago Plan between 1906 and 1908. The Plan's railroad section in turn spurred the formulation of six or more plans for the reform of the city's passenger and freight operations.[52] The basic Chicago Plan noted:

> Year by year the railroads have gone on straightening their lines, reducing grades, and building additional tracks; and the result has been large savings in operating expenses. The time has now come to devise some plan whereby the enormous terminal costs will be

lessened materially; and that city will benefit most wherein this problem shall be worked out first and best.[53]

Many Chicago businessmen complained that air pollution interfered with their operations. The treasures of the Art Institute were in danger of serious damage. The Illinois Central Railroad, which had right of way that forestalled attempts to beautify the lakeshore and adjoining park system stemming from the example of the Fair was a particularly sore point for politicians and planners, but all thirty-nine railroads and switching companies in the city were under heavy pressure to reduce their emissions for health and economic reasons. (Comparable pressure was being put on railroad operations in all the major US cities.) Public Health officers pointed with pride to the decline in average annual deaths by twelve per 100,000 population from impure water diseases such as typhoid following reform of the water and sewage systems. They pointed with alarm to the corresponding increase of twenty-two deaths per 100,000 population from impure air diseases such as tuberculosis, pneumonia, and bronchitis which they linked to air pollution.[54] Moreover, the world of air pollution control was aware of relationships between acid rain and plant growth, between generally polluted air and the economic costs of wasted fuel, added cleaning and material maintainance, and increased depreciation. London at the turn of the century was said to suffer added costs from air pollution of about £5.5 million or some $6 *per capita*. A study in Cleveland, Ohio concluded that its added costs in 1909 were at least $12 *per capita* and possibly twice that amount. A more elaborate study in Pittsburgh, Pennsylvania finished in 1913 set its minimum added costs from polluted air, excluding health effects, at $20 per capita, at a time when typical working class incomes were about $600 per year.[55]

By about 1907, a coalition of engineers, planners, public health officials, businessmen, and progressive politicians was forged to force the railroads into their next stage of progress, the electrification of the Chicago terminal system, the largest and most complex system in the world. Electrification was seen as making a large contribution to the pollution abatement efforts, but the abaters recognized that it could not be justified economically unless the overall system were consolidated and simplified. Correspondingly, each of the reorganization schemes of the planners presupposed electrification as one of the keys to increased efficiency of operation and the release of surplus real estate necessary to offset some of the capital expenditure and permit further development of the central business district. They saw the potential for smoke abatement as a valuable adjunct to the creation of a planned City Beautiful.

Moving as fast as the various bureaucracies could respond, there was a flurry of activity in the City Council and the Association of Commerce. Reports from the Council Committee on local transportation and the City's Chief Smoke Inspector set the railroad's share of the local smoke at forty-three per cent and the cost of polluted air at three-fourths that of Cleveland, or $8 *per capita* for a total of about $18 million per year.⁵⁶ Electrification was judged to be technically feasible and, in the case of the Illinois Central, cost beneficial. If the cost figures for the Illinois Central could be generalized as the Transportation Committee believed, everyone would benefit from progress. By the autumn of 1909 a local ordinance had been drafted requiring all railroad trains within eight miles of city hall to be propelled by electrical power. The City Council found that they could not legislate any one particular remedy and a year later rewrote the ordinance under consideration to indicate that all railroad trains or cars operated within seven miles of the Court House after January 1, 1913 shall be operated by power other than steam or in a manner so as not to produce smoke or other noxious gases that injuriously affect the public health, comfort or convenience.⁵⁷

The Association of Commerce did not stand idly by in the face of proposed ordinances with such far reaching implications. It formed an internal committee of seven experts to consider the matter.⁵⁸ The committee included Paul Bird (the Chief Smoke Inspector), Bion Arnold, and William Goss who had recently been named Dean of Engineering at the University of Illinois. Professor Charles Merriam, regarded as the founder of the behavorial school of political science and a prominent progressive Republican politician, was also on the committee.⁵⁹ Their report was generally favourable to electrification providing it was allowed to take place over a reasonable time. The Committee of Seven suggested that the work could start with the suburban and other passenger services of several of railroads which had a dense traffic, and eventually be extended to all passenger, freight, and switching services within the city limits. They did not examine the cost in detail although they understood that it would run to several hundred million dollars. Their attitude on costs was expressed as follows:

> The advantages of improvements, such as Track elevation in Chicago, the adoption of air brakes and automatic couplers, were recognized by many railway officials, but their general adoption was reluctantly accepted owing to the financial difficulties involved. Air brakes and automatic couplers were, however, found to be appliances of such value in the handling of trains as to now make it appear strange that their adoption should ever have been opposed. ... it was found that they were of large economic advantage to the railroads, making not only the operation of trains more convenient and safe, but also adding to the capacity of their trackage and equipment. Similarly, where electrification has been adopted by steam railways, greatly increased capacity has resulted
>
> Track elevation in Chicago was brought about to satisfy the public demand for reduction in the number of casualties at grade crossings. ... It has been found, however, that track elevation has been of sufficient economic advantage to the railroads to justify the investment required for its creation; thus, judging from all precedents, it is reasonably certain that electrification ... will

prove to be of economic value to the railroads, and also that ... its application will necessarily be brought about by pressure of public demand.[60]

The Committees of Seven went on to call for other reforms on the smoke abatement scene, but they had already gone too far. Marvin Hughitt, president of the Chicago and Northwestern Railroad, spoke out at a private meeting of representatives of the Association and railroad presidents in September 1910:

> If it costs $100,000,000 or $400,000,000 to electrify the terminals of Chicago, the question is, will the commerce of Chicago bear the burden? ... Commerce must pay; you businessmen must pay.
>
> I feel it is premature to be seriously considering the electrification of the terminals of the railroads in Chicago. It is premature to put such a burden upon commerce.[61]

The point was well taken. Indeed some Association interests had assumed that the committee would 'quiet the popular clamor against the railroads'.[62] Whatever the merits of electrification, if allowed to evolve naturally at the option of the railroads, the section of the report suggesting that the change would have to be forced as in the case of air brakes and couplers would scarcely 'quiet' the agitation. The executive committee of the Association decided against releasing the report. They indicated uncertainty about the financial feasibility of the project and pressed for a full investigation. Typically, a high-level Commission was established — the Chicago Association of Commerce Committee of Investigation on Smoke Abatement and Electrification of Railway Terminals. The mayor and the railroads named four members each, while the Association named nine members. The railroads paid for the investigation on an open-ended cost basis which ran to $650,000 over the next four and a half years. The Chief Engineer of the investigation was paid $36,000 a year!

The Committee of Investigation evidently had *carte blanche*. Two matters of critical importance were settled in its early meetings — the scope of the investigation and the territory to be investigated. The scope comprised the following articles:

Article I.   A determination as to the necessity of changing the motive power of steam roads to electricity or other power.
Article II. The Mechanical or Technical Feasibility of such change.
Article III. The Financial Practicability of any such change.[63]

An elaboration of Article I is given in Figure 4 to indicate the incredible detail typical of the investigation.[64] The art of Technology Assessment was fashioned at the highest level sixty years ago. At the suggestion of its Chief Engineer, Horace G. Burt, the Committee of Investigation accepted the limits of the entire Chicago switching district as the territory to be studied. This switching district included all 194 square

SAFETY		
Passengers	Liability to personal injury	On trains On tracks On right-of-way At stations
Employees	Liability to personal injury	On trains, locomotives or cars On tracks On right-of-way At stations At engine-houses, shops, power-stations, etc.
Public	Liability to personal injury	On tracks On right-of-way At stations At grade-crossings At other places
HEALTH		
Passengers Employees Public	Liability to injury by	Smoke Gas Cinders and dirt Noise
COMFORT and CONVENIENCE		
Passengers Employees Public	Affected by	Noise Smoke Gas Cinders and dirt Time
LOSS and DAMAGE		
Passengers	Personal property on trains by	Storms Derailments
Employees	Personal property on trains and right-of-way	Collisions Wrecks Fire
Public	(a) Real estate (buildings, parks, land) (b) Personal property (store stocks, household furnishings, decorations and clothing, business, vegetation)	Smoke Gas Cinders and dirt Noise
Note: Railway Companies' loss will be considered when investigating Financial Practicability of the change and will include	(a) Baggage and express freight, including livestock (b) Maintenance of equipment (rolling stock, shops, machinery and tools, power-plants, etc.) (c) Fuel (d) Maintenance of way and structures (tracks, bridges, buildings, etc.)	

Figure 4. Scope of investigation on smoke abatement and electrification of railway terminals in Chicago.

miles of the City as Zone A; and 236 square miles outside the city as Zone B, of which some forty-three square miles was in the industrial area of northwest Indiana. Burt correctly anticipated that the city Council would soon demand that electrification extend to the city limits and he argued with respect to Zone B:

> This territory while beyond the jurisdiction of the City of Chicago, is so closely related thereto industrially and commercially as to practically form an integral part of the territory under consideration, and for that reason has been included in the Scope of Investigation.[65]

The district contained about 4000 miles of track and required an average of nearly 1700 locomotives in daily operation pulling the equivalent of about 440 miles of cars. Forty per cent of the mileage was running track with the rest in sidings, yards, freight and passenger terminals, and in private industrial tracks. None of the critics stated the complexity of the situation any better than Burt who noted in part, 'the terminals within Zone A contain not less than one hundred and five separate yards and as many junction points of intersecting tracks of which an enormous volume of through and local traffic is daily interchanged and handled'.[66] The planners, who were denounced editorially in a railroad journal as being unconcerned with costs, saw such complexity as a mandate for consolidation, simplification, and electrification.[67] 'Burt wished to establish the gigantic character of any such undertaking. The decision to include the whole switching district and to exclude the issue of consolidation from the investigation were critical in shaping the outcome. Later, Goss, who became Chief Engineer upon the death of Burt, maintained that it was the City that had insisted 'upon the complete and immediate electrification of its steam railway terminals as they now exist'.[68] The known records indicate otherwise. Goss went on to urge the establishment of a separate Commission having large powers to study in conjunction with the railroads the whole terminal problem. He concluded:

> The working out of any large scheme of terminal betterment, which can be justified as an economic measure, would at once provide an opening through which electrification as a detail could be, and probably would be, readily admitted.[69]

A City Terminal Commission was established to study the overall terminal problem under the direction of John Wallace, a former general manager of the Illinois Central and the controversial former chief engineer of the Panama Canal. A rival Citizen's Terminal Commission was established under the leadership of Bion Arnold to check on the work of Wallace. Later, Arnold joined forces with Wallace in a continuation of the work of the City's Commission.[70] (It should be noted that Wallace was paid at the same rate of $3,000 per month as the chief engineer of the Association's Committee of Investigation.) These various Commissions could agree on no more than piecemeal reform, but they all assumed that electrification was inevitable and that the two issues of electrification and smoke abatement were so interlaced that they should be carefully considered in conjunction with each other. And yet, the issues were neatly separated, which probably was what some people wanted. The Committee of Investigation frustrated everyone with its interminable delays. The three-year-old report of the Committee of Seven was leaked to the City Council in May 1913 in an apparent attempt to undercut the Committee. The Investigation had an enormous undertaking, however, and it would not be hurried no matter how angry the City Council might become.

The Committee of Investigation operated with a staff of full-time experts and specialized consultants, and was aided by advisory committees on various technical matters. These experts were drawn largely from the railroads, but they included those with experience in actual terminal electrification projects. Thus, the Investigation used the services of George Gibbs of the New York City firm of Gibbs and Hill who was chief engineer of electrical traction and station construction for the Pennsylvania Terminal scheme then nearing completion. Another noteworthy expert was the consulting chemist William Hoskins of the Chicago firm of Mariner and Hoskins who did much to innovate air-quality sampling techniques. After a thorough search of the literature, Hoskins and his workers adopted or improved upon the best sampling techniques then available.[71] Using a mobile unit, they achieved a profusion of sophisticated measurements that were probably not surpassed for the next forty years. Hoskins's stack-sampling techniques were superb extensions of ideas developed by John Aitken and John Owens in Great Britain, although it should be noted that such techniques are still evolving today in the search for truly representative samples (see Figure 5).

Determined to get to the core of the matter the Investigation monitored the flow of coal to different classes of service in great detail. An unprecedentedly detailed emission inventory was prepared for all sources of pollution except coke and gas plants: heavy metal industries and other manufacturing processes; steam ships and tug boats; high pressure boilers in large buildings and power plants; low pressure boilers or coal furnaces in households and small buildings; the grit from unpaved streets, refuse disposal, construction, and other activities; as well as from locomotives and stationary boilers in the railroad yards (see Figure 6(a) and (b)). To drive home to the various commerce and civic interests that major polluters existed beyond the railroads and heavy industry, all large boiler stacks were plotted on a map and the buildings of a typical downtown business block were depicted stripped to their smoke stacks to reveal the profile of a modern industrial complex, as in Figure 7.

The Investigation also noted that the smoke from the 25,000 registered motor vehicles, which consumed one per cent of the energy

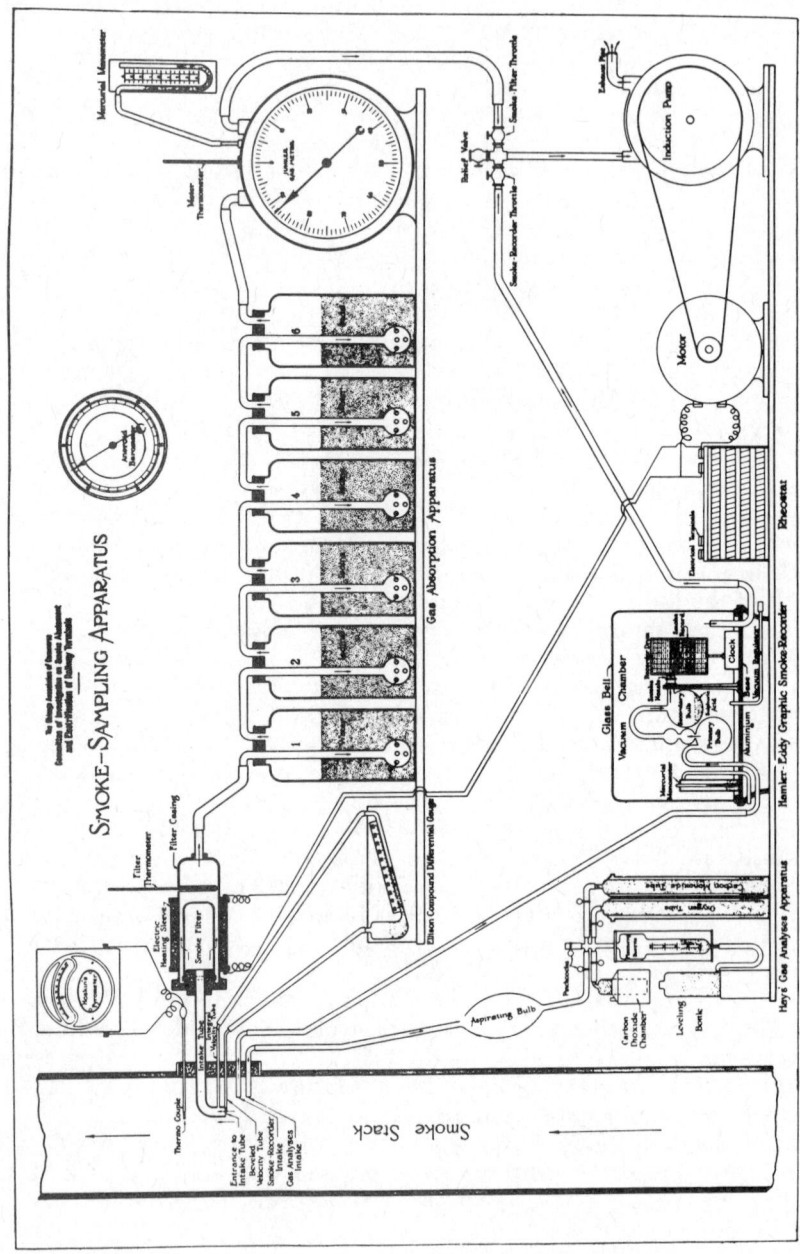

Figure 5. Flue gas sampling apparatus as connected to a stationary stack under test.

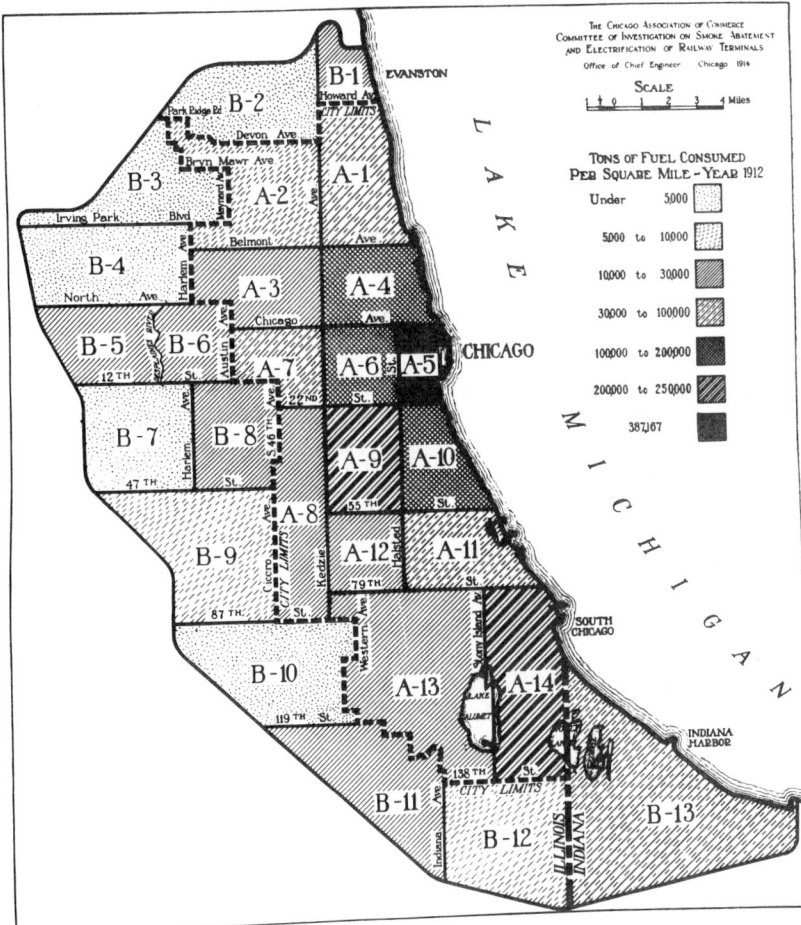

Figure 6(a). Relative density of fuel consumption in the several zone districts of the Chicago area.

total, already constituted a significant source of pollution that would continue to grow. It is ironic that the mobile sampling unit was housed in a gasoline-driven automobile because its speed, power, and range were greater than an electric automobile and because its engine could be shut off during sampling to avoid any contamination.[72] The electric automobile would have continued to give off bothersome levels of emissions from its large array of batteries. The emissions from the battery necessary to power the air sampling equipment were unavoidable in any event and evidently sufficiently small to be ignored.

The fuel requirements and emissions for a typical locomotive burning the smoky high-volatile, high-ash, and high-sulphur coal of

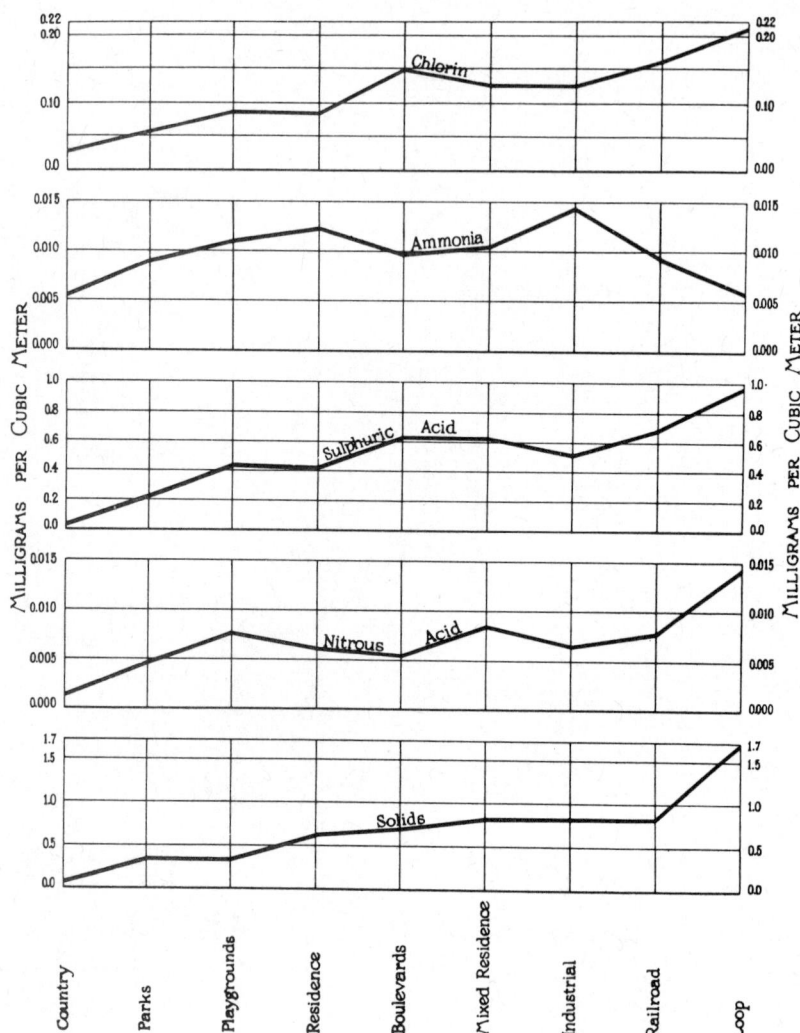

Figure 6(b). Average results of the analyses of samples of air.

Illinois and Indiana that supplied the railroad systems of Chicago were determined at the Altoona test facility of the Pennsylvania Railroad equipped with dynamometer, spark hood, and smoke collector. Goss had pioneered many of these techniques in the 1890's at Purdue University and in 1904 for the Pennsylvania Railroad test facility at the St. Louis World Fair.[73]

The Investigation concluded that the steam locomotive was vastly overrated as a producer of pollution in the context of all the other

Figure 7. A typical downtown Chicago business block stripped of its high rise office tower shells to reveal its true character as a pollution source rivalling that of railroad facilities and factory complexes. The numbers indicate high pressure power plant smoke stacks.

sources, including fugitive or allied sources from inadequate municipal housekeeping practices. Rather than contributing forty per cent to fifty per cent of the air pollution, as was commonly asserted, a flood of measurements were produced to show that in the City, Zone A, locomotives produced only twenty-two per cent of the smoke, seven per cent of the larger particles, and ten per cent of the gases (see Figure 8). It was thought that the sulphur emissions were most deleterious. It is significant that the emitted volumes of carbon dioxide, carbon

150   *Two Problems in Fuel Technology*

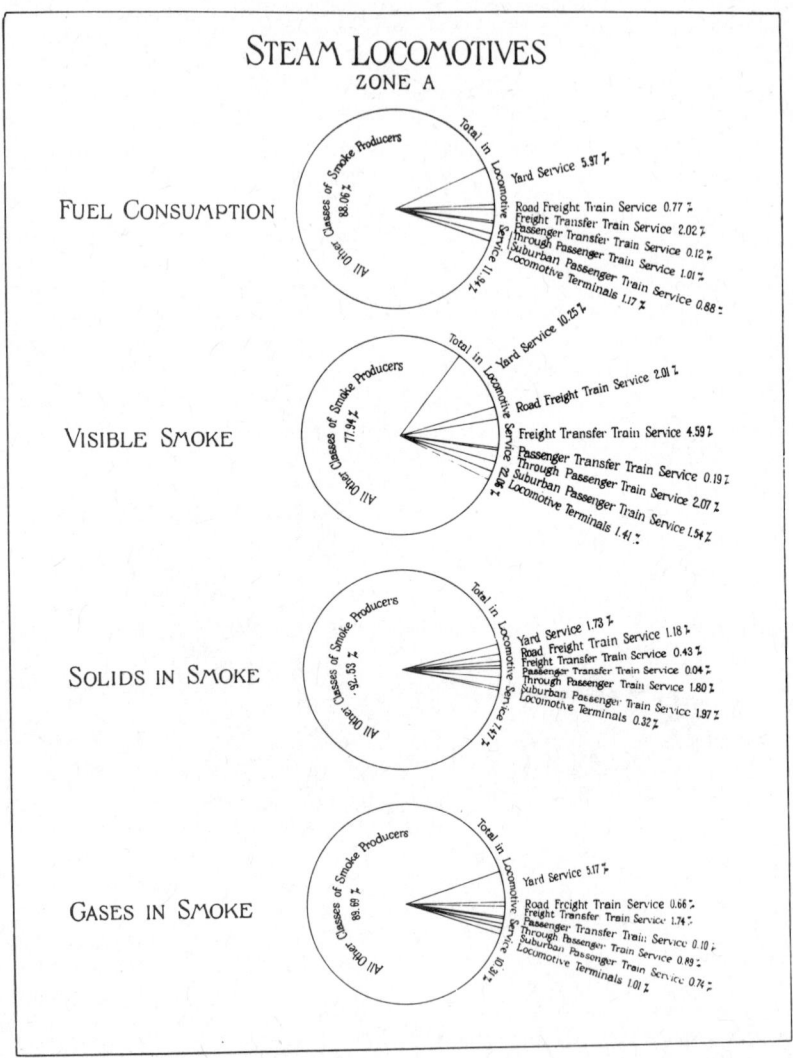

Figure 8. The relative importance of steam locomotives in Chicago as consumers of fuel and producers of pollution.

monoxide, sulphur dioxide, sulphur trioxide, and various nitrogen compounds were combined for the purposes of these calculations. The Investigation did not tackle the still vexing task of trying to assign relative significance to the various quantities of gases.

The Committee reviewed the technologies, costs, and circumstances of all existing electrification projects. A team headed by George Gibbs visited Rudolph Diesel's test facility in Switzerland where a heavy-duty locomotive was under development. They also reveiwed other forms of traction power including batteries and compressed air. They balanced one set of occupational hazards against another. Above all, using results tabulated on a punch card system containing details of every movement on sample days, they gave extremely detailed attention to the power and cost requirements of the changeover under different electrical design options. They included in the cost estimates details overlooked in earlier evaluations, such as the cost of a telephone system to report a fault before the entire system was jeopardized. They included the cost of abandoned capital (down to the last ash pit), the 'precipitated costs' of advancing the date of implementing longer-term improvements that would have to be done in any circumstances, and the added labour costs, under unchanged work rules, of operating shorter outlying division points with more deadhead time within the terminal. They reviewed existing legal restrictions against financial co-operation between the railroads and local governments and new restrictions by the Interstate Commerce Commission on handling capital costs. In short they generated an exhaustive account of the tangible costs of change with no attempt at consolidation or simplification. It is significant that the Investigation was unwilling to suggest any pressure on the trade-union Brotherhoods, the ICC, or state and city government to facilitate the project, for that would have been stepping outside the bounds of their self-generated charge.

The Committee of Investigation concluded that the choice definitely was between steam and central electric power. Thomas Edison had heaped scorn on the backwardness of steam operation and the concomitant pollution during a visit in 1912.[74] In the Committee's judgment, his promise to news reporters of effective low-cost battery powered locomotives could not be fulfilled. They indicated that, in time, the diesel engine might provide a viable third way for main line express service, but even Diesel himself, just before his death, expressed doubt that such a locomotive could be adopted for the critical start and stop switching operations which were shown to cause about one-half of the terminal smoke problem. Moreover, the report noted that the diesel engine would produce smoke under some conditions of operation and so represented an unknown tradeoff in pollution. Ironically, with the clamour for cleaner air as part of the stimulus, the first diesel switch engine with an electrical transmission went into experimental service within about seven years, with diesel-electric locomotives for passenger and freight operations coming on line in the 1930's.

As for electricity, there were many design problems without precedents and the total effort if carried out would have more than doubled the existing world experience in passenger electrical traction. Nevertheless, the project was judged to be technologically feasible. The Committee's inventiveness and resourcefulness in preparing the report might have been evidence enough that almost any technical problem could be overcome. Yet the project could not be recommended because of the prohibitive capital costs and the minimal impact on air quality.

Cost figures were developed for each of three systems; 600-volt direct current third rail, 2,400-volt direct current and 11,000-volt alternating current overhead contact systems. When adjusted to the growth expected by 1922, the capital cost of the cheapest system, 11,000-volt A.C., was estimated to be $188 million. The net salvage value of current capital was set at $10 million, and the precipitated cost of advanced capital improvements was set at $96 million, for a net increase of $274 million. Added interest charges of five per cent and added depreciation charges of four per cent or five per cent (depending on the facility), would increase the fixed costs by $17 million a year without considering added taxes or precipitated costs. Operating costs would be expected to decrease, largely through savings in coal by only $2.3 million a year. The report gave no hint of current costs, but tentative calculations indicate that they may have been in the order of $100 million, with the net increase of $15 million or about fifteen per cent.[75]

The improvement in air quality was judged on the basis of 1912 relationships and measurements. The inventory indicated that 21 million tons of coal and coke were burned during the year in the proposed electrification zone, or about nine tons *per capita*. Of this total, 0.5 million tons were burned by railroads in stationary boilers and 2.8 million tons were burned in locomotives. A central electric power plant to be located in the middle of the city and operating at an efficiency of 2.5 pounds per kilowatt-hour or a heat rate of about 2900 BTU per kilowatt-hour would require only 0.8 million tons to provide equivalent power for traction. The projected fuel saving of electric over steam power would thus be 2.0 million tons per year, but, since some of the power would be delivered outside of the city, the net reduction within the city would be only 1.4 million tons or about eight per cent. This net reduction coupled with the emission characteristics of a modern power plant was calculated to lead to a decrease averaged over the entire city of about twenty per cent in smoke and about five per cent in solids and gases. The designation of 'smokeless' for the large well-regulated fires of a central power plant can be appreciated in these figures, but an increase of fly ash, for example, would offset some of the other savings that might follow the reduction in fuel consumption. At any rate, the overall improvement was alleged to be too small to be noticed, which was true only if the reductions were averaged over the entire city and the effects on particular neighbourhoods were not considered. Nevertheless, the study clearly showed how much had to be done before the whole City could enjoy very much better air quality.

After an unsuccessful attempt to gather information through a circular letter, the Committee decided not to press for an investigation into the economic costs of polluted air or to make what is now called a cost-benefit analysis. If they had, they undoubtedly would have asserted that the total added costs from all sources of air pollution were in the neighbourhood of $20 million of which no more than perhaps ten per cent would have been saved by electrification against total added costs to the railroads of $15 million.

But their earlier study recognized many intangibles, such as the cost of pollution in terms of health and quality of the railroad service. In a nutshell (and it should be remembered that nutshells float on surfaces), the Committee concluded, after extended discussion, that the problem was much worse in other cities. They argued that even in the most polluted cities the effects were detrimental but not disastrous to health. Moreover, in Chicago most of the harm, indeterminant as it was, came from sources other than the railroads. On the matter of quality of service, they noted that electrification would increase convenience and comfort for users and neighbourhoods alike in the following respects:

> By permitting a greater celerity of train movements
> By increasing the reliability of train movements
> By reducing localized smoke sources
> By reducing noise.[76]

But the noise issue was ignored. The issue of localized smoke was blunted, as has been noted, by averaging all local sources over the entire city. The increased speed and reliability of movement would have increased capacity, but these issues were handled by implying that greater capacity could be obtained by degrees as necessary and at a far cheaper capital cost.

The financial condition of many of the railroads throughout the States was precarious during the second decade of this century.[77] Overall, railroads invested something like $4.8 billion in the decade between 1907 and 1916, with generally flat or decreased net earnings as a reward. The railroad and investment journals published numerous articles on the declining productivity of capital under regulation and the increase of operating costs as a result of excessive capitalization. The Interstate Commerce Commission was slow in granting rate increases. Significantly higher wages were being mandated by other Federal action. Direct government involvement was not seen as much help in the light of the vast over-runs experienced in the construction of the Canadian National Transcontinental.

Investment analysts estimated that new railroad capital to provide proper growth nationwide was $1.25 billion a year. This needed capital would have to be financed by the ordinary money markets. In 1916, the railroads were able to raise only $64 million, almost exclusively through the sale of bonds; a stock offering by the powerful New York Central, for example, was regarded as a complete failure. As early as 1912, it had been argued:

> If the general public will recognise the extent of these capital requirements ... there need be no doubt that the unwisdom of discouraging such investigations by excessive, irritating, and hampering regulations will be perceived and that wise and conservative counsels will prevail.[78]

The needs for terminal electrification in Chicago were small in comparison to the sums needed nationally. Nevertheless, the financial pinch was felt and several of the companies operating in Chicago were in very poor financial condition. A requirement for them to abandon much of their existing capital and enter an electrification scheme would have been regarded by some people as little less than a confiscation of property and a possible threat to future investment.

The railroad critics used similar information to draw conclusions that might justify the project. As stated earlier, the investigation did not address itself to the issues of land use, congestion, and general reorganization and re-construction. Analysis of these factors might have altered the financial conclusion. One letter-writer reminded the public that the steam railways were the greatest barrier against business expansion. Alfred Beirly suggested that if the yards and terminals in the heart of the city were relocated one mile to the south, land with a potential value of at least one billion dollars would be released and would permit a four-fold expansion of the central business district.[79] The estimate may have been high by a factor of two or three, but a net increase in land value of only one or two hundred million dollars would have done much to make the project financially practical.[80] Neither side seems to have mentioned the anti-trust decision against the consolidated terminal in St. Louis, so perhaps that precedent posed no threat.[81] In any event, the problems of determining the details of co-operation equitably between the railroads would have been very difficult. If the project had been realized in Chicago, however, it is clear that an alphabet of cities such as Atlanta, Buffalo, Cleveland, and Denver would have promulgated their own electrification schemes.

The Committee of Investigation argued that the national character of the railroads, the need for equal treatment in the city and across the nation, and many other extenuating circumstances pointed to an entirely different attack on the problem. Stressing the importance of a 'broad view', without however defining the 'better interests', they asserted:

> ... objections to smoke are valid only in so far as they may be based upon conditions which are avoidable, and ill-advised restrictions against the use of fuel are not only unreasonable, but harmful to the larger and better interest of the community. . . . The problem of smoke abatement cannot be solved by any simple campaign for the enforcement of restrictions ... It is a living question, since it assumes new aspects as the state of the art develops.[82]

The Committee suggested that progress is something required of and expected from all sectors of the economic household. Smoke inspection records confirmed that great progress had already been achieved. Yet, they noted that every user of coal or oil could achieve still greater efficiency of use and produce less nuisance in the process. The technology of steam locomotion, for example, was not at a standstill, and there were known expedients, such as larger grates and the simple fire arch, that could be used more widely to promote better combustion.

The same was true of other classes of energy use — including, of course, electric power and traction. Scrubbers and electrostatic precipitators were being developed. What was needed locally was a step-up in the ongoing activities of the city's smoke abatement department. The regulation and inspection of all sources of pollution, an education campaign based upon the best exemplars, and technological change would bring the problem progressively under better control. Angus Smith, the first Chief Alkali Inspector in Great Britain, could have spoken from the grave that such a reasonable approach was not likely to succeed.

The results of the committee's work were paraded before a number of technical societies where anticipated advocates of electrification either reversed their earlier positions or remained silent, except for one electrical utility official who noted wryly that if the charge had been worded differently, the conclusion might have been different. In any case, resumés of the work were given decent burials in the technical literature where recognition of the social costs issues was virtually nil and where the work was soon forgotten.[83]

The City Council, backed by cries of outrage from the local press, had tried to keep the proposal alive.[84] The real issues were fought behind closed doors where the notion of an influential Councillor John Wallace that, although electrification was desirable and still inevitable it would have to come piecemeal, prevailed. John Wallace returned to the strategy of trying to force his old company, the Illinois Central, to agree to the immediate electrification of its suburban service and the construction of a new lakefront terminal. This new terminal would in time be electrified and thereafter attract the passenger service of other companies in some sort of co-operative arrangement. Presumably, electrification would spread slowly to the remaining operations. Many members of the full council continued to press for the larger proposal and demanded time and time again that the Council Committee on Railway Terminals issue a report so that action could be taken. Finally in 1919, with the political retirement of Dr. Willis Nance (a prime mover of the larger scheme and chairman of the Council's Health Committee), a report appeared, the issue was discussed and killed. Within a few months agreement was achieved between the Illinois Central and the City on the establishment of an electrified suburban service, the building of a new terminal, and the eventual electrification of all its locomotives within the city.[85]

Wallace did not live to see any of the results. A few months after issuing the final report, he died on a trip to Washington where he was going to testify before the ICC. The electrified suburban service was in fact established, but the Lakefront Terminal was never built and eventually diesel-electric locomotives were accepted as fulfilling the Illinois Central's agreement to electrify. The business district expanded slowly to the north of the Loop, with the river proving to be less of a barrier to it than the railroad properties to the south. (More recently, the first, fifty-year-old rolling stock of the suburban venture went on the market for sale to traction buffs. When last seen, the old main line station was an empty, weed-filled lot.)

Meanwhile the drive toward railroad electrification had stayed alive nationally. For example, it was discussed at a joint meeting of the American Society of Mechanical Engineers and the American Institute of Electrical Engineers in 1920. Also, the concept was a main feature of the 'superpower Project', which led to a survey in 1920-21 by the United States Geological Survey of the Boston to Washington corridor studying the possibility of widespread electrification.[86] A. H. Armstrong reached the vice presidency of the General Electric Company on the basis of his advocacy of the new technology, but eventually he was forced out due to the disappointing sales.[87]

A recent report by the Federal Railroad Administration notes, 'railway electrification in the United States dates back more than 70 years. It reached its peak during the 1930's but with the advent of diesel-electric locomotives, it ceased to expand and has since declined from a peak of over 2500 route miles to less than 1200 route miles today'.[88] The report goes on to note five basic problems which the early projects were meant to solve; namely, air pollution and noise, capability for high-speed and acceleration in passenger service, interconnection of passenger terminals for greater efficiency and greater convenience to passengers, increased track capacity and related operating-efficiency factors, and increased efficiency in mountainous terrain. The first four of these problems were at issue in Chicago. That ill-fated scheme alone would have added over 3,000 electrified miles to the national system. The issues are far from dead. As noted in the current Federal report, '... a new look at the role of rail electrification is appropriate because of changing energy costs, advances in electric and catenary technology, and possible changes in the traffic patterns of the rail system'.[89] Whether the relative shift in political and economic power, the reality of current energy and capital markets, and availability of massive government subsidies will make a difference remains to be seen. The technology assessments of today have a long way to go to reach the sophistication, and perhaps the sophistry, of this nearly-forgotten early effort.

Notes

Part 1 of this paper is a portion of a study funded by the Center for Urban Studies, Wayne State University, U.S.A. Part 2 is an expanded version of the paper 'Pollution and Progress' read at the Midwest Junto of the History of Science Society, March 1976. Figures 1, 2 and 3 were drawn by the author. Figure 4 has been based on an illustration in The Chicago Association of Commerce Committee of Investigation on Smoke Abatement and Electrification of Railway Terminals, William F.M. Goss, Chief Engineer, *Smoke Abatement and Electrification of Railway Terminals in Chicago*, Chicago, 1915.

1. M. King Hubbert, *U.S. Energy Resources*, Senate Committee on Interior and Insular Affairs, Serial No. 93-40 (92-75), United States Government Printing Office (G.P.O.), June 1974, pp. 196-197.

2. *Scientific American*, XII (June 27, 1857), p. 239.
3. *Wall Street Journal*, October 28, 1975, p. 11.
4. For Gramm see *Wall Street Jounral*, November 30, 1973, p. 8, and *New York Times*, April 27, 1974, p. 37. For Lund see *Detroit Free Press*, February 6, 1975, p. 3A. Also see *Forbes*, May 15, 1974, p. 116. ('the whale oil hypothesis').
5. *Wall Street Journal*, November 30, 1973, p. 8.
6. David B. Brooks and P.W. Andrews, 'Mineral Resources, Economic Growth, and World Population', *Science*, Vol. 185 (July 5, 1974), p. 14.
7. The best introductory sources include: [Editors of the Pyne Press], *Lamps and Other Lighting Devices 1850-1906*, Princeton, 1972; Matthew Luckiesh, *Artificial Light*, New York, 1920, esp. Ch. XVI; Edgar W. Martin, *The Standard of Living in 1860*, Chicago, 1942, pp. 94-97, *passim*.; Centennial Seminar, *Oil's First Century*, Harvard Business School, 1960; and Harold F. Williamson and Arnold R. Daum, *The American Petroleum Industry, The Age of Illumination 1860-1895*, Volume I, Evanston, Ill., 1959. A Canadian view, with much information relevant to the situation in the United States, is Loris S. Russell, *A Heritage of Light: lamps and lighting in the early Canadian home*, Toronto, 1968.
8. Gramm, *Wall Street Journal*, Note 4.
9. The best general introduction to the whaling industry and my source of data on supplies and prices is Walter S. Tower, *A History of the American Whale Fishery*, Philadelphia, 1907. Also useful is L. Harrison Matthews, 'A Note on Whaling', *A History of Technology*, eds. Charles Singer and others, New York and London, Vol. I, pp. 55-63.
10. Prices from Tower, *The American Whale Fishery*, Note 9; the Federal Reserve Bank of New York's concept of cost of living was followed, from *Historical Statistics of the United States, Colonial Times to 1957*, G.P.O., pp. 111 & 127.
11. Tower, *The American Whale Fishery*, p. 8, Note 9.
12. *Historical Statistics*, Series A181-209, Note 10.
13. A carcel mechanical lamp with 7/8 inch wick produced 7.5 candlepower burning sperm oil at the rate of 2 fl. oz./hr. or about 16 gallons for 1000 hours, according to Benjamin Silliman, Jr., *Report on the Rock Oil, or Petroleum from Venago Co., Pennsylvania*, New Haven, 1855, pp. 19-20 (a convenient reprint is in J.T. Henry, *The Early and Later History of Petroleum*, Philadelphia, 1873 (reprinted, New York, 1965) pp. 52-53.)
14. Assuming the spermaceti produced was twenty per cent of the sperm oil: two million gallons of spermaceti at 7.5 pounds per gallon yield 15 million pounds of candles; burned at the rate of two candles every other night, 1/6 pound per night or 60 pounds per year per household equals 250,000 households. Sperm whales were said to yield an average 60 barrels of oil (*Encyclopedia Britannica*, 8th ed., Vol. 24, p. 528) while an ordinary sized sperm whale was said to yield 12 large barrels of crude spermaceti, density 9.433 (*McCulloch's Commercial Dictionary*, article, 'Spermaceti'). The point needs further substantiation.
15. Emerson McMillin, 'American Gas Interests', *One Hundred Years of American Commerce*, Chauncey M. Depew (ed.), 1895 (reprinted New York, 1968), Vol. 1, p. 295.
16. *Encyclopedia Britannica*, 8th Ed., article, 'Philadelphia'.
17. McMillin, *op. cit.*, I, p. 296, Note 15.
18. *Ibid.*, I, p. 297, Note 15.
19. Louis Stotz and Alexander Jamison, *History of the Gas Industry*, New York, 1938, pp. 9-10, *passim*. Harold C. Passer, *Electrical Manufacturers 1875-1900*, Cambridge, Mass., 1953, reminds us about how much still needs to be determined about the gas light industry.
20. In 1859, E. Merriam tallied 83 deaths, 106 serious injuries, and $44,000 in property loss that could be attributed to camphene and similar burning fluids. The editor of *Scientific American* had warned repeatedly of the hazards and noted this time, 'we long ago ordered this stuff out of our house, and we advise all our readers to do the same thing. Use coal oil, tallow candles, pine knots, anything rather than hazard life, limb, and property by the constant use of a dangerous fluid'. *Scientific American*, new series II (Jan. 2, 1860) p. 51. Also see issues for VI (Oct. 5, 1850), p. 24; VIII (Dec. 11, 1852), p. 101; XI (Aug. 2, 1856), p. 374; and sources listed in note 7 above.

21. *1905 Census of Manufactures*, p. clxxii.
22. *Scientific American*: for gas, II (Feb. 20, 1847), p. 147; for lard oil, VI (Aug. 23, 1851), p. 386; for camphene, XI (Aug. 2, 1856), p. 374.
23. Victor S. Clark, *History of Manufactures in the United States*, New York, 1929, Vol. I, pp. 492-4; *1860 Census of Manufactures*, *passim*.
24. Quoted in *Scientific American*, V (Feb. 16, 1850), p. 176.
25. J.T. Henry, *History of Petroleum*, p. 138, Note 13.
26. Quoted without credit in R.J. Forbes, *More Studies in Early Petroleum History 1860-1880*, Leiden, Netherlands, 1959, p. 143.
27. *Scientific American*, XI (June 7, 1856), p. 312.
28. The pioneering reappraisal of the beginning of this industry in the United States is Kendall Beaton, 'Dr. Gesner's Kerosene: The Start of American Oil Refining', *Business History Review*, XXIX (March 1955), pp. 28-53. See also *Oil's First Century*, Note 7, and Williamson and Daum *American Petroleum Industry*, Note 7.
29. *Scientific American*, new series II (June 2, 1860), p. 3.
30. See Benjamin Silliman, Jr., *Rock Oil*, Note 13; Abraham Gesner, *A Practical Treatise in Coal, Petroleum, and other Distilled Oils*, 2nd ed., New York, 1865 (reprinted, 1968), pp. 123-4; Forbes, *Early Petroleum History*, pp. 130-3, Note 26; and James C. Booth and Thomas H. Garrett, 'Experiments on Illumination with Mineral Oils', *Journal of the Franklin Institute*, 3rd, series XLIII (June 1862), pp. 373-80.
31. Bureau of Labor Statistics, *History of Wages in the United States from Colonial Times to 1928*, revised, *Bulletin 604*, G.P.O., 1934, p. 254, etc.
32. Williamson and Daum, *American Petroleum Industry*, p. 81, Note 7.
33. Williamson and Daum, *ibid.*, pp. 105, 107, and 110, Note 7; Beaton 'Gesner', pp. 47-8, 5507, 159, Note 28; and Henry, *History of Petroleum*, pp. 191, 277-8, Note 13. There is a vast popular and technical literature on library shelves treating post World War I fears of a petroleum famine and the endeavours to develop Colorado oil shale formations. During this period Henry Ford funded research and development to produce alcohol from potatoes to make farmers independent of the greedy oil barons. The economic numbers were never quite right, however, and, despite sporadic attempts since 1930, any residual hope for large scale development of any of these materials was swamped by the rather belated discovery of the East Texas oil field.
34. *New York Times*, April 5, 1976, p. 36; April 19, 1976, p. 32. The debate over the origins of these 'delights', as in Craig Claiborne's recent articles, needs to shift to the more fruitful ground of popular acceptance. Similarly, as it comes of age, the history of technology needs to supplement the usual concerns about discovery, invention, and innovation with studies about the factors that foster or hinder technological change.
35. Hamlin Garland, *A Son of the Middle Border*, New York, 1917, reprinted 1962, pp. 389-91; Henry Adams, *The Education of Henry Adams*, Boston, pp. 379-90, reprinted in Thomas Parke Hughes, *Changing Attitudes Toward American Technology*, New York, 1975, pp. 168-175.
36. Samuel P. Hayes, *Conservation and the Gospel of Efficiency: The Progressive Conservative Movement 1890-1920*, Harvard University Press, 1959, reprinted 1969, pp. 2-5.
37. See especially the speeches of Samuel Insull collected in two privately printed volumes, *Central-Station Electric Service*, Chicago, 1915 and *Public Utilities in Modern Life*, Chicago, 1924.
38. William F.M. Goss, 'Steps in the Development of a Smokeless City', *Proceedings of the Indiana Academy of Science*, Indianapolis, 1906, pp. 75-6. Biographical sketch in *Who's Who in Engineering*, 2nd. ed., 1925.
39. James Hatch, 'Development of the Electric Railway', *Journal of the Western Society of Engineers*, 13 (1908), p. 504; paper read at the Society meeting, February 14, 1904.
40. Bion J. Arnold, *Journal of the Western Society of Engineers*, 12 (1907) p. 12; paper read at American Institute of Electric Engineers meetings in Chicago, September 1904.
41. A useful study emphasizing the experience in Cincinnati, Cleveland, Pittsburgh, and St. Louis is Robert D. Grinder, 'The Anti Smoke Crusades; Early Attempts to Reform the Urban Environment', unpublished PhD dissertation, University of Missouri, Decem-

ber 1973, University Microfilms 74-18, 539. The best contemporary short book length treatments, from a British point of view, are William Nicholson, *Smoke Abatement; a manual for the use of manufacturers, inspectors, medical officers of health, engineers, and others*, London, 1905 and the masterful analysis of the situation in Leeds, England by Julius B. Cohen and Arthur G. Ruston, *Smoke: A Study of Town Air*, London, 1912. The Chicago report on smoke abatement released in 1915, see Note 63, provides the most comprehensive single contemporary review of the subject. It needs to be considered with the *Mellon Institute Smoke Investigation Bulletins, No. 1-10*, Pittsburgh, 1913-22.

42. Scrapbooks of the Pittsburgh smoke investigation, 3 vols. Consist of roughly 10,000 clippings from nearly 200 newspapers for the years 1912-13. Now located at the Carnegie Library of Pittsburgh.

43. Horace Spencer Fiske, *Chicago in Picture and Poetry*, Chicago, 1903, p. 6.

44. *Ibid.*

45. D.H. Burnham and E.H. Bennett, *Plan of Chicago*, The Commercial Club, Chicago, 1909.

46. Booth Tarkington, *The Magnificent Ambersons*, New York, 1918, Avon Library reprint, 1967, p. 7. Also see Chapter XXVIII, especially pp. 193-6. This Pulitzer Prize winning work was Tarkington's second attempt at writing the urban novel.

47. Carl W. Condit, *Chicago 1910-29: Building, Planning, and Urban Technology*, University of Chicago Press, 1973. Condit provides a masterful analysis of this period of growth and a convenient summary of the Chicago terminal situation. Also see *Plan of Chicago, op. cit.*, Chapter V, and Harold Melvin Mayer, 'The Railway Pattern of Metropolitan Chicago', PhD dissertation, University of Chicago, 1943. Other details from *Electrical Traction*, 7 (1911), pp. 478-9; 8 (1912), p. 659; 9 (1913), p. 444.

48. Booth Tarkington, *The Turmoil*, New York, 1915, p. 1. Serialized in *Harper's Magazine*, August 1914 to March 1915. Republished as part of the triology *Growth*, 1927 which also included *The Magnificent Ambersons*, Note 46, and *The Midlanders*, 1923.

49. *Plan of Chicago*, Chapter III, Note 45.

50. Stephen Longstreet, *Chicago 1860-1919*, New York, 1973, p. 160.

51. *Plan of Chicago*, p. 62, Note 45; Frederic A. Delano, 'The Chicago Plan, with Particular Reference to the Railway Terminal Problem', *Journal of Political Economy*, XXI (November 1913), 819-31.

52. City Club, *A Series of Addresses before the City Club dealing with the problem of Reorganizing Chicago's Railway Terminals*, June 3-10, 1913; George E. Hooker, *Through Routes for Chicago's Steam Railroads*, City Club of Chicago, 1914.

53. *Plan of Chicago*, p. 62, Note 45.

54. W.A. Evans, 'The Harm of Smoke', pp. 22-7 in Foreman, see Note 56.

55. John J. O'Connor. Jr., *The Economic Cost of the Smoke Nuisance of Pittsburgh, Mellon Smoke Investigation Bulletin No. 4*, Pittsburgh, Pa., 1913. London, p. 7; Cleveland, pp. 8-10; Pittsburgh, p. 43.

56. Chicago, City Council, Committee on Local Transportation, Milton J. Foreman, Chairman, *The Electrification of Railway Terminals as a Cure for the Locomotive Smoke Evil in Chicago, with Special Consideration of the Illinois Central*, Chicago, 1908; Chicago, Smoke Inspection Department, Paul P. Bird, Smoke Inspector, *Report (first), 1909-1910*, Chicago, 1911.

57. *Journal of the Proceedings of the City Council of the City of Chicago*, October 19, 1908, p. 1551; September 27, 1909, p. 1204; December 19, 1910, p. 3289; February 6, 1911, p. 3725. Apparently, the change in the distance from the centre of the city was intended to simplify the problem of electrification by excluding two major junctions just over seven miles out from the centre toward the south and the northwest.

58. The Committee had eight members, but W.L. Abbott of the Commonwealth Edison Company did not participate. It was referred to as the Ewen Committee after the chairman, John M. Ewen, or the committee of seven. The role of the Edison company in the dispute has not been traced. Samuel Insull, see Note 37, and his lieutenant, Peter Junkersfield, see Note 87 below, strongly supported the idea with or without the co-operation of the utility.

59. Barry D. Karl, *Charles E. Merriam and the Study of Politics*, University of Chicago Press, 1974, p. viii. Merriam went on to become an alderman on the city council,

but failed in his attempts to become mayor. His political frustrations represent the decline of the scientific programme of progressive conservative reformation.

60. Chicago Association of Commerce Report, dated July 8, 1910. Printed in reference 30, pp. 19-23. Also see, Christopher Clark, 'The Railroad Safety Problem in the United States, 1900-1930', *Transport History*, 7 (Summer 1974), pp. 97-123.

61. Hooker, *Chicago's Steam Railroads*, p. 25, Note 52.

62. Proceedings of the Railway Terminals Committee, City Council of Chicago, June 2, 1913, pp. 19-20. See also minutes for May 12, 19, 26, and 29. Copies of 'Smoke Papers' at John Crerar Library, Chicago, see Note 64.

63. The Chicago Association of Commerce Committee of Investigation on Smoke Abatement and Electrification of Railway Terminals, William F.M. Goss, Chief Engineer, *Smoke Abatement and Electrification of Railway Terminals in Chicago*, Chicago, 1915, p. 1046. See next Note.

64. 'Scope of Investigation on Smoke Abatement and Electrification of Railway Terminals', Committee archives, V, #3. The archives or 'Smoke Papers' located at John Crerar Library, Chicago, comprise five vols. totalling some 2500 pages, mostly mimeographed. Included are articles or abstracts from the literature search, minutes, progress reports, tentative outlines, speeches, and other documents.

65. 'Remarks by Mr. Horace G. Burt, Chief Engineer', Committee Archives, I, #20.

66. *Ibid.*

67. Editorial, 'Electrification and the Smoke Nuisance', *The Railway Engineering Review*, 48 (May, 1908), p. 369. Also see articles, 'The Railroad Electrification Question in Chicago', and 'Enough Legislation Against Railroads', pp. 370-2.

68. W.F.M. Goss, 'Concerning the Electrification of Steam Railway Terminals', Address before the City Club of Chicago, June 3, 1912, Committee Archives, II, 17, p. 7. The point is pivotal. The time-table allowed within the proposed ordinances may have been much too tight to permit significant rearrangement in the critical downtown area. Presumably the matter could have been negotiated, although the Committee of Investigation never showed any such inclination.

69. *Ibid.*

70. *Report of Mr. John F. Wallace to the Committee of Railway Terminals of the City Council of Chicago*, Oct. 20, 1913; Bion J. Arnold, *Report on the Re-arrangement and Development of the Steam Railroad Terminals . . .*, Nov. 18, 1913; *Preliminary Report of the Chicago Railway Terminals Commission*, Mar. 29, 1915, see esp. pp. vi-vii, 20-1; *Report of John F. Wallace, Chairman, Chicago Railway Terminal Commission*, March 1921, p. 10

71. Benjamin Linsky, 'Appraisal', in abridged ed. of air pollution sections of the basic 1915 report of Committee of Investigation (see Note 63), Maxwell Reprint, 1971. Linsky is Professor of Civil Engineering, West Virginia University (retired), and former air pollution control officer in Detroit and San Francisco. He notes, 'the investigators knew quite well what they were looking for, and where and how to find it. The technical apparatus used in studying the air pollution situation was as modern as it could have been, with perhaps a number of innovations for the time'. Further work is in progress on these points and on the Committee's assessment of the meteorological controls of air pollution.

72. 'Brief Outline of Methods which Commend Themselves for Adoption in the Investigation of the Atmosphere of Chicago', Committee Archives, I, #22 (Mar. 27, 1912), p. 1.

73. [W.F.M. Goss], *Locomotive Testing Plant*, Purdue University, LaFayette, Indiana, 1895; Pennsylvania Railroad, *Locomotive Tests and Exhibits*, St. Louis, 1904.

74. *Chicago Evening Post*, Jan. 6, 1912; *Chicago Tribune* editorial, Jan. 8, 1912 (unpaged news clips).

75. Unpublished Ms. by author, 'On the Cost of the Chicago Switching District'.

76. *Smoke Abatement and Electrification of Railway Terminals in Chicago*, p. 1027, Note 63.

77. John F. Stover, *The Life and Decline of the American Railroad*, Oxford University Press, 1970; Pierpont V. Davis, 'The Railroad Situation', an address before the Transportation Club of Louisville, Ky., April, 1917. Articles on 'Railways', in *The New*

International Year Book, New York, Dodd, Mead and Co. for years 1910 to 1916 provide very useful summaries. Also see articles on 'Electric Railways', and 'Smoke Abatement'.

78. H.T. Newcomb, 'Railway Capitalization and Traffic', *Moody's Magazine*, April 1912, in Committee Archives, IV, # 21, p. 7.

79. Chicago Daily Journal, December 9, 1915, p. 18b. Beirly worked on the reform of the Chicago Harbor Terminals, as did Merriam (Note 59) and several other principals in this study.

80. Homer Hoyt, *One Hundred Years of Land Values in Chicago, The Relationship of the Growth of Chicago to the Rise in its Land Values, 1830-1933*, University of Chicago Press, 1933, pp. 221, 260; and, 'Excerpt from Report of Board of Economics and Engineering to Mr. S. Davies Warfield, President, National Association of Owners of Railroad Securities, Inc. on Joint Terminals at Chicago, Illinois', New York, March 28, 1922.

81. *The New International Year Book for 1912*, New York, 1913, p. 607.

82. *Smoke Abatement and Electrification of Railway Terminals in Chicago*, p. 285, Note 63.

83. *New York Railroad Club Proceedings for 1915-1916*, pp. 4379-4460; *Journal of the Western Society of Engineers*, 21 (April 1916), pp. 310-29; *Journal of the Franklin Institute*, March 1916, pp. 305-38; *Railway Mechanical Engineer*, 90 (Jan. 1916), pp. 3, 11-14; *Railway Review*, 57 (Dec. 4, 1915), pp. 722-34; *Engineering News*, 74 (Nov. 25, Dec. 9, Dec. 16, 1915), pp. 1024, 1141, 1186; *Chicago Commerce*, XI (Dec. 3, 1915), pp. 12-33, 40.

84. *Chicago Daily Tribune*, Nov 4, 1915, p. 6; Dec 2, pp. 1, 6; Dec 3, p. 8; Dec 7, p. 15; Dec 9, p. 6. *Chicago Daily News*, Nov 24, 1915, p. 15; Nov 30, p. 3; Dec 2, ?; Dec 4, p. 8; Dec 7, p. 4; Dec 8, p, 8. *Chicago Daily Journal*, Dec 2, 1915, pp. 6, 18; Dec 9, p. 18. *Chicago Evening Post*, Dec 2, 1915, p. 11. *Chicago Herald*, Dec 2, 1915, p. 3; Dec 3, p. 6. The issue was before the City Council on 10 occasions between Dec 6, 1915 and May 19, 1919 when it was finally 'filed'.

85. John Wallace, *Report of John F. Wallace, Chairman, Chicago Railway Terminal Commission*, March 1921. Wallace died July 3, 1921. Biographical sketch in *Dictionary of American Biography*, X, 1936.

86. *Railway Age*, 69 (1920), pp. 327, 553, 610, 732-3, 739-46, 1115; 70 (1921), pp. 727-8.

87. George Wise, personal communication. Armstrong's publications include, 'Electricity as Applied to Steam Railroads', *The Railway and Engineering Reivew*, 45 (1905); 'Conservation of Railway Resources', *Electric Traction*, 13 (June 1917), pp. 423-36; *The Future of Our Railways*, General Electric Co., 1920; 'The Economic Aspects of Railway Electrification', *Journal of the Franklin Institute*, 191 (April 1921), pp. 493-502. Also see Samuel Insull, Notes 37 and 58, and Peter Junkersfield, 'Electric Service Problems and Possibilities', *Proceedings of the Engineers' Society of Western Pennsylvania*, 32 (1916), pp. 61-120.

88. Taskforce on Railroad Electrification, William E. Loftus, Chairman, 'A Review of Factors Influencing Railroad Electrification', Federal Railroad Administration Report No. FRA-OPP-74-01, Feb, 20, 1974, p. 1.

89. *Ibid.*, p. 2.

The Historical Development of Theories of Wave-Calming using Oil

JOHN C. SCOTT

INTRODUCTION

Although it is certain that the calming effect of oil poured on waves has been appreciated for many centuries, knowledge regarding the causes of this effect has progressed only very slowly since the first attempted explanations in classical times. Real advances in the theory of the subject have occurred most irregularly and unevenly, and even today there is neither an adequate explanation nor even a widespread appreciation among scientists of what has and has not been established. For a subject which in the eighteenth and nineteenth centuries aroused much interest, both scientific and popular, and which was then and is now of vital interest to a considerable section of the community, this may be thought surprising.

Several factors appear to have been important in the slow development of the subject: the problem is undoubtedly complex even from a modern scientific point of view, involving several imperfectly understood topics; the wave-calming phenomenon itself has two almost distinct aspects, whose relationship is still not quite clear, even though one of them can now be considered to be well understood; direct observation of valuable calming effects in dangerous breaking seas is not accessible to most people; and experiment, either at sea or in the laboratory, is beset by serious difficulties.

Scientifically, the subject involves the interaction of a turbulent air flow with a complex non-linear surface-wave field, and this interaction itself is not understood in sufficient detail to allow the consideration of the effect on it of adding an oil layer to the water surface. However, we do now have the advantage over scientists of earlier times in understanding many of the important effects of adding oil to water surfaces, because of the advances in surface chemistry that have been made in the past 100 years. These advances have enabled us to identify two separate, though not completely distinct, aspects of the wave-calming phenomenon. Oil and other surface-active substances have a marked damping action on waves generated on still water in the absence of wind. This aspect of the action of oil is easily observable in the laboratory and is now quite well understood, following the pioneering work of Osborne Reynolds, Lord Rayleigh, and Horace Lamb at the end of the last century, and the work of V.G. Levich, R. Dorrestein, and many others in the years since 1940.

The factor that is of crucial importance for this damping effect, the surface dilational elasticity,[1] is also likely to be of major significance, both for the effect of oil on the generation of waves when wind blows on an initially calm surface and for the tendency of large waves to break. These two latter effects of oil appear to be intimately related, since it seems likely that it is the generation of small waves on the relatively smooth backs of large waves that precipitates their breaking action. However, the relation of these two effects with the damping of progressive waves on still water is less certain, even though the same property of the oil film — the dilational elasticity — may well be responsible for all of the important observed phenomena. Perhaps this dual nature of the calming effect has been instrumental in causing some of the confusion which is evident even in modern writings.

The restriction of the observation of oil behaviour on dangerous seas to the relatively small proportion of the population who put to sea in vulnerable vessels appears also to have been an important factor in the neglect of the subject. Although several hundred experienced sailors have gone on record, particularly in the latter quarter of the nineteenth century, with their belief in the efficacy of wave oiling, many people remain sceptical. Only a well controlled scientific experiment, making meaningful measurements, will finally convince the scientific world, and this experiment is still conspicuously lacking. As will be seen, the closest attempt to fulfil this need was made by the Scottish engineer John Shields, with his experiments in Scottish harbours in the 1880s. It is arguable, however, that since these observations were purely visual, and to some extent subjective, they could have been open to error.

It is the purpose of this paper to describe the progress of theories of the subject to date, and to show the evolution of our present picture of the mechanisms involved in the phenomena.

Antiquity

The phenomenon of the calming of waves on water by the application of an oil layer — 'pouring oil on troubled waters' — is one of the earliest to appear in literature, and it is probable that the effect had been in practical use for centuries before the first authors — Aristotle, Plutarch, and Pliny the Elder — came to record it. Even had they not known of it from earlier seafaring peoples, it would have been most unlikely if the Greeks, with their extensive seagoing activities and their far-ranging trade in one of the most effective wave-calming materials, olive oil, had not discovered the effect independently.

From these earliest writings, two aspects of the phenomenon are apparent. The most widely known of these, even if it is still not universally accepted among scientists, is the effect of oil on storm waves when poured from a distressed vessel. Plutarch, in his *Natural Phenomena*, ascribes to Aristotle the belief that the role of oil is in causing a smoothness of the surface, so that the wind can make no impression and raise no swell.[2]

The second recorded effect of oil — the more frequently recorded in the classical works — concerns an apparent effect of the oil on the light passing through the surface into deep water. This effect is implicit in the *Problems* of Aristotle,[3] in Plutarch's *The principle of cold*,[4] and in Pliny's *Natural History*,[5] and as indicated by Plutarch, the observations are almost certainly associated with the practice of Mediterranean divers, in both ancient and modern times, of releasing oil beneath the water so that when it rises to the surface it spreads and improves the light available for fishing.

Although some of the eighteenth and nineteenth century authors have assumed that this purported improvement in light was to do with the optical properties of the air/water interface, it is unlikely that such simple optical effects would be significant in practice. Barger, Garrett, Mollo-Christensen, and Ruggles recently measured the light intensity beneath the sea surface, including observations under an oil patch, and they found no significant improvement in intensity.[6] It is much more likely that the beneficial effect reported in this case is concerned with steadying of the light — the removal of the more distracting components of the flicker associated with the passage of light through the undulating, rippled surface. If this is so, then the effect of the oil on the transmitted light is caused by suppression of wind-generated surface disturbances, and may therefore be considered to be more closely related to the effect on storm waves.

Of the few fragments that have reached us concerning the opinions of classical authors on the effects of oil on water there is little today that we can fully understand. Our conception of the nature of matter differs so fundamentally from theirs that our limited understanding of the terms they used, even assuming as accurate a translation as possible, would preclude any detailed consideration of their ideas. There is however enough that rings true in the theory attributed by Plutarch to Aristotle to indicate that someone of authority, even if it was not Aristotle himself, had early appreciated the beginnings of an explanation.[3] This fragmentary theory, which even today forms the basis of our understanding of the phenomenon, is that the oil produces calm by smoothing the water surface so that the wind can slip over it without making an impression. While this idea may, perhaps fairly, be considered merely to change the question, 'How does oil calm the waves?', into 'How does the oil smooth the surface?', it also will appear to us to represent an approach much more familiar to the scientist than is the type of obscure explanation more usual in classical authors, based on apparently vaguely defined inherent properties of natural materials such as sea, air, and oil.

The importance we may ascribe to Aristotle's suggestion will simply depend on whether we see the words 'smoothing the surface' as another way of saying 'removing the waves'. or whether we credit Aristotle with perceiving that the smoothing of the surface of a large wave will reduce the force exerted on it by the wind, and so reduce the chance of its breaking. It is obviously not possible to make a definitive judgement.

After the classical period, as would be expected, the subject appears to have been neglected, although we still find occasional brief references. Theophilus Simocrate, the sixth century Byzantine author, mentioned the effect in the form of a dialogue, and even offered an explanation:

> The wind is a very subtle and nimble thing, Polycrates, and oil is sticky, greasy, and shiny, So the wind, because of the polished effect of the sea, slides over the surface, and cannot gather up waves upon the waters.[7]

The Venerable Bede, in his *Ecclesiastical History of the English People*, completed in A.D. 731, described a miracle of St. Aidan in which holy oil was given to sailors setting out on a voyage, for use if a storm should arise.[8] It is reported that the desired effect was produced, the sea ceasing its fury. This tale was retold in the seventeenth century by Henricus Canisius in his *Antiquae Lectionis*,[9] and by Simon Majolus, Bishop of Volturara, in his *Dies Caniculares*.[10] Majolus shows concern lest such 'miracles' should appear to detract from the miracle of Jesus Christ in stilling a storm without the need for oil (Matthew, chapter 8, Mark, chapter 4). In his *Colloquia*, Erasmus includes a mention of the use of oil on waves, again in the popular dialogue style.[11]

The eighteenth century

The first clear published account of the phenomenon is due to Benjamin Franklin, who described several instances of the popular usage of wave oiling, reported a series of experiments he made himself with ripples on ponds, and also attempted an explanation.[12] While it can well be argued that his contribution to the theory of the effect was hardly greater than may generously be attributed to Aristotle (and his contemporaries, especially Frans van Lelyveld,[13] and Martin Wall,[14] were not slow to point this out), a perusal of the number of papers which were written as a direct result of Franklin's work, and a consideration of the interest it obviously generated throughout the Europe of his day, emphasizes the considerable debt owed by science to this part-time scientist.

Franklin does not in fact claim to have discovered the effect, as indeed he cites many instances of its common usage,[15] and references are made both to Pliny and to an item in Thomas Pennant's *British Zoology*.[16] It is also true however, as was pointed out by Wall, that his grasp of the concepts of adhesion and cohesion, by which he explains the spreading of oil on water, is not very sure. Regarding spreading, Franklin says:

> But if there be a mutual repulsion between the particles of oil, and no attraction between oil and water, oil dropt on water will not be held together by adhesion to the spot whereon it falls; it will not be imbibed by the water; it will be at liberty to expand itself; . . . and the expansion will continue, till the mutual repulsion between the particles of oil is weakened and reduced to nothing by their distance.' Note 12, p. 453.

His explanation of the smoothing action is similarly suspect:

> Now I imagine that the wind blowing over the water thus covered with a film of oil, cannot easily catch upon it, so as to raise the first wrinkles, but slides over it, and leaves it as smooth as it finds it. It moves a little the oil indeed, which, being between it and the water, serves it to slide with, and prevents friction, as oil does between those parts of a machine, that would otherwise rub hard together'. Note 12, p. 453.

To our scientific thinking, the idea of reducing the friction between two fluids by interposing a more viscous third fluid is not reasonable, although it appears to have required the practical demonstrations of John Aitken in 1883[17] and of R.N. Ivanov in 1937[18] finally to discredit this hypothesis.

Franklin's contribution was therefore not principally in the insight which he brought to the scientific explanation of the phenomenon. It was in the fact that he introduced indisputable knowledge of the subject to the scientific world for the first time, applying the principles of observation, experiment, and theoretical construction in a new area. The correctness of his deductions was a relatively small factor in exciting the minds of his generation, and it is arguable that if he had been more correct he would have had less success. Coming to the subject from what could loosely be called the field of international affairs at a time when the New World was just beginning to make its presence felt, he managed to stimulate nautical men such as Frans van Lelyveld[13] and Captain J.G.D. Müller,[19] ecclesiastics such as L'Abbé Mann[20] and Paolo Frisi,[21] and academics such as Albrecht Meister[22] and Martin Wall.[14] Such an explosion of interest was not to occur again until the last quarter of the nineteenth century, when Shields' experiments in harbours, experiments in the laboratory by Plateau, Marangoni, van der Mensbrugghe, and Aitken, and another century of nautical experience, together resulted in a comparable rush of papers, pamphlets, and even books on the subject.

In the years immediately following publication of Franklin's paper, however, there were no serious challenges to the friction-reduction theory he had proposed. Frans van Lelyveld, in his pamphlet published in 1775, both in Dutch and in French, added greatly to the general interest in the subject by offering a prize of 30 ducats for the best set of answers to a series of questions concerning the means of most effective use of the effect, and reported experiments, although he did not show much interest in its correct explanation. It is not known if this prize was ever awarded, but the date for applications was extended for a year in the 1776 edition of his work. Van Lelyveld reports that the East India Company instructed its captains to make experiments whenever the opportunity arose, and to make a report of their findings.

The response of the established church to the interest in this subject, once in the category of a miracle, appears to have been simply to deny the existence of the effect itself. This at least was the response of Paolo Frisi, the Barnabite, who proposed that the appearance of the phenomenon was the result of observers being deluded by the superficial smoothing of the surface, and that in reality the sea was not diminished. Frisi exhorted modern scientists to admit that they might have made a mistake, in the same way that earlier scientists had frequently erred in their search for the truth. A second sceptic, Joannes Le Francq van Berkhey, published a brochure about this time, vigorously attacking the suggestions made by van Lelyveld, apparently on the grounds that reducing the waves in one region must make the sea more dangerous elsewhere.[23]

Sceptics were however a minority, and once the seeds had been sown by Franklin and van Lelyveld, interest in the effect grew steadily for some years. The seeds do not appear to have prospered everywhere in Europe, however, and the majority of contributions to the discussion at that time appeared in Germany, following the contributions of Müller and Achard. This German contribution appears to have had little impact outside Germany itself, and even there, written discussion had effectively ended well before the matter was reviewed by Weber and Weber in their 1825 textbook on waves.

L'Abbé Mann,[20] who in 1780 published in Belgium the results of experiments in which he observed the effect of a wide variety of oils on several open bodies of water, offered an explanation that is significantly different from the 'lubrication' explanation of Franklin. He considered that the generation of waves by wind was a result of the considerable natural affinity between air and water, evidenced by the tendency of water to evaporate into the air — a tendency which increases in the open air and in the presence of a cold wind. The effect of oil was therefore simply in separating the two fluids. Since air had no comparable affinity for oil, waves could not be excited in the same way. At a time when phlogiston theory was still in vogue, this type of explanation apparently seemed not unreasonable. The natural affinity of water for air would explain why the water appeared to *want* to be rushed along by the current of air. The idea took a long time to fade away completely, and as recently as 1893, a Scottish engineer, W.J. Millar, was proposing a similar explanation for the reduction by oil of the 'friction' between air and water.[24]

In 1778, Franz Carl Achard, later to develop the first sucessful process of extracting sugar from sugar beets, described what appear to have been the first truly laboratory-scale experiments on wave-breaking, although he failed to recognize the role of the wind in the practical application of wave-oiling.[25] He examined the effect of oil on mechanically generated standing waves in a rectangular tank of considerable size, 4.3m × 1.2m × 1.2m, and observed the tendency of these waves to sink a model ship of length 150mm. Using as a criterion the time required for the model to fill with water and sink to the bottom, he showed that oil had an appreciable, though not remarkable, calming effect on the waves, and he further showed that attaching floating solid bodies around the ship might better reduce the effect of breaking waves. Achard believed that the calming effect both of oil and of floating solid bodies was due to the load they imposed on the surface, giving an inertial resistance to the lifting up of water necessary for wave formation. In experiments using olive oil and fennel oil, he observed that uniform layers were not formed, the oil preferring to accumulate in droplets on the surface, and it is almost certain that he considered that the parts of the surface not covered by these droplets were not covered by oil at all. The greatest objection to his experiments, however, is that he ignored completely the role of the wind in the breaking of waves at sea and even though he mentions Franklin's work, it is therefore likely that he had not read his paper.

Achard's work attracted the attention of several interested observers. In his paper of 1782, Captain J.G.D. Müller expressed his doubt about the relevance of Achard's model-scale experiments and indicated that whereas the efficacy of oil-pouring was to him an established fact, he was not so sure about Achard's suggestion of using large floating sealed casks moored to the ship.[19] He probably felt, as did Albert van Beek in 1841,[26] that such casks might be thrown against the ship by the waves, thus causing more damage than the waves themselves.

Captain Müller believed implicitly in the great value of wave-oiling for reducing hazards at sea and he suggested that oil was effective because of its viscosity, considerably greater than that of water. He felt that it was obvious that the application of a viscous skin to the surface of water simply did not allow the violent motions necessary for wave breaking, even though he conceded that oil might have no sensible effect on the amplitude of a swell. He likened theories of affecting the swell itself by action at the surface to hypotheses regarding the construction of pyramids starting at the top!

Johann Friedrich Wilhelm Otto, writing in 1798, favoured the type of explanation offered by Mann (1780), that the oil separates the air from the water and prevents their natural vigorous interaction.[27] He also, however, accepted Achard's reasoning in the use of large floating boxes, although he felt that oil should be the more effective, being lighter and able to spread further. Professor Friedrich Christian Kries, of Gotha, wrote in 1799, critically comparing Otto's paper with that of Müller, which Otto had apparently not read at that time, and showing scepticism regarding all of the wave-damping mechanisms that had so far been proposed.[28] He felt that the 'chemical affinity' explanation, originated by Mann, was likely to be less significant than mechanical causes, bearing in mind the forces associated with fluctuating winds, but he did not reject it entirely. With Müller, he dismissed Achard's work as being of little relevance.

Kries was least opposed to the 'viscous skin' concept advanced by Müller, although he doubted whether a viscous film could be strong enough to prevent the breaking of a wave. He felt that more experiments were required to test Müller's theory.

The beginnings of wave theory

At the end of the eighteenth century, the subject of wave-oiling appeared to be developing quickly. Contributions had been made by a number of learned men, and there were several theories available, as well as a few experimental results. The subject even appears to have made some impact on the non-scientific community, for the poet Schiller, in his 'Hero and Leander', of 1801, wrote:

> Alle Göttinen der Tiefe,
> Alle Götter in der Höh'
> Fleht sie, lindernd Öl zu giessen
> In die sturmbewegte See

It is therefore suprising that interest seems to have declined just at the time at the beginning of the nineteenth century when the hydrodynamics of wave motion was beginning to be established. The work of Gerstner and others was becoming known,[29] and the concept of circular-orbit particle motion was increasingly used in accounts of the subject. However, for the first quarter of the nineteenth century, the subject of wave-oiling made comparatively little advance, and interest appears to have been on the wane. One fresh theory was put forward, by Robinet in 1807,[30] and although Robinet himself did not feel that it was important enough to allow it more than a brief mention at the end of his paper, it is a theory which, with diminishing plausibility, tends to recur throughout the nineteenth century.

Robinet was principally concerned with a somewhat different observation made by Franklin in his 1774 paper, that a vessel containing a thick layer of oil on top of water, when it set into vibration, could show violent oscillations of the oil/water interface, even though the oil surface above it might be relatively still. In attempting to explain the two phenomena, Robinet assumed that they were connected, and he explained wave-damping by proposing that the thin covering of oil on the water surface flowed in such a way that it filled the wind-induced wrinkles, presenting a smooth upper surface to the wind. The possible objection to this idea that the layer was not thick enough to fill the size of wrinkles commonly observed on a wind-blown water surface, he countered by observing that all ripples must start from nothing and that the effect of the oil must therefore take place when they are still sufficiently small.

Baron Franz Xaver de Zach published an account in 1822 in which he indicated his approval of Mann's hypothesis that wave generation was a result of the air content of water being below saturation.[31] He quotes a member of the Royal Humane Society:

> Je crois que l'une des causes du vent est lorsque l'eau n'est pas suffisamment saturée d'air. Un courant de ce dernier fluide se précipite alors vers l'eau, comme vers un vacuum.

In support of this hypothesis, several rather strange observations are described which concern apparent differences in the weather conditions prevailing simultaneously over land and sea.

The next major advance in wave-oiling theory occurred in 1825, when the brothers Weber, Ernst Heinrich and Wilhelm, published the first textbook on wave theory, *Wellenlehre, auf experimente gegründet*.[32] This book presented wave theory in a generally accessible form and it included a considerable section on the phenomenon of wave-oiling. The authors' view of the suppression of wind-generated ripples was that the oil layer did not properly adhere to the water surface, and that the horizontal component of the wind merely acted to push the oil layer along without applying any stress to the underlying water. Although this idea derives essentially from the original Franklin concept of friction reduction, the Webers' work is evidence of a more realistic approach to the problem, recognizing that oil cannot act in the same way as it does between a pair of sliding solid surfaces. The Webers' own experimental observations told them that oil had only a slight effect on the propagation of waves generated by dropping stones into the water, even though they found that a layer of lead oxide formed on the surface of liquid mercury had a marked damping effect on waves generated in this way. They acknowledged a form of elasticity of the surface covered by oil, as evidenced by the formation of interference colours when the film becomes stretched by a rising bubble. Nevertheless, they supported the Franklin friction-reducing theory without question, and spent some time explaining how in their view the small vertical component of the wind that remained after the oil had rendered the horizontal component ineffective should not be capable of generating waves.

At this time, the concept of surface tension was almost unknown, and oil was supposed to exert only hydrostatic forces on the water surface. As we approach the middle of the nineteenth century, we encounter the early stages of the theory of cohesion and surface tension. However, although M.L. Frankenheim in his pioneering book, *Die Lehre von der Cohäsion*,[33] in 1835 realized the importance of cohesion in the spreading properties of oil films, he nevertheless favoured the hydrostatic explanation of wave damping, originally proposed by

Robinet, that oil flows from crest to trough, eliminating the waves. This view was criticized in 1858, as surface chemistry began to acquire momentum, by Paul du Bois-Reymond, in one of the earliest extensive experimental studies of the spreading properties of liquids.[34] He felt that since oil films were effective in damping even when they were thin enough to give optical interference colours, it was not reasonable to suggest that there was sufficient oil to flow into wave troughs. He returned to Franklin's friction-reduction hypothesis, but also suggested that the water-particle trajectories associated with the waves might be so distorted by the surface flow caused by the oil film as it spread that wave motion might thereby be disrupted. He recognized that this mechanism could be valid only for a short time, at the beginning of oil pouring, and he found it necessary to assume the correctness of Franklin's theory in order to explain the considerable persistence of wave damping.

The Commission of the Royal Netherlands Institute, 1842

In the 1830s and 1840s the Dutch, who appear to have made no contribution until this time since the work of van Lelyveld, burst into activity. Paulus van Griethuyzen published a book in 1834, which reviewed the subject and proposed the use of wave-stilling phenomenon for the protection of Holland's complex and vulnerable system of dykes and sea walls during storms.[35] He discusses the project in some detail and also reviews earlier work without, however, analyzing the mechanisms which had been proposed in explanation of the effect. P. de Leeuw commented critically on van Griethuyzen's suggestion in a letter published in 1837 — part of the occasional evidence we have of an undercurrent, not merely of disbelief in the wave-stilling phenomenon but of firm belief in the harmful effects of oil-pouring.[36] De Leeuw believed that pouring oil in one location actually made the sea more dangerous in others, and it is possible that this view was quite widely held.

Albert van Beek, a member of the Royal Netherlands Institute, wrote a paper in Dutch in 1841 giving a detailed review of earlier research, and urging the thorough investigation and employment of the phenomenon.[26] The following year he lectured on the subject to the French Academy in Paris, and a French translation of his paper soon followed. He appears not to have read to works of Müller, Otto, and Kries, although he was acquainted with the work of Achard, which he too considered misguided. He favoured the type of 'natural affinity' explanation of Mann.[20] In support of this hypothesis, he quotes the observations of one Mr. J. Boelen, a captain of the Dutch navy, who claimed that the sea rises to a greater height during fine, calm weather that it does when the sky is overcast and the air misty. From this rather dubious observation, much in the same character as those reported by de Zach, van Beek concludes that dry air exercises more adherence than humid air, and that complete separation of air and water using an oil layer will result in the surface remaining free of waves.

Van Beek appears to have had some influence in the Royal Netherlands Institute, since a Commission was appointed as a result of his proposal, consisting of three Members of the First Class of the Institute, to investigate the truth behind the claims. This first Commission unfortunately failed to agree, with apparently only one Member, A. Lipkens, giving any credence to the effect, and a second Commission of five Members was charged with carrying out suitable experiments. This second Commission reported, rather surprisingly, that there was no value at all in the phenomenon for protecting sea-defences,[37] and from the evidence which is still available, we see that this conclusion triggered a storm of controversy.

The Commission had chosen to perform their experiments in June 1841 at Zandvoort on the Dutch coast, and as their critics were able to point out, tempests are not common there at that time of year. Despairing of a suitable storm, the members had chosen a day with a moderate wind, and unfortunately, with that wind blowing parallel with the coast. As was soon made clear both by Lipkens and van Beek in public protests at the decision, oil is not expected to have a large effect under such conditions and the experiments were therefore irrelevant.[38] Nevertheless, van Beek's cause had been damaged, and the adverse conclusion of the Commission may well have been an important cause of the neglect of the subject during the subsequent forty years.

The development of theories involving surface tension

The concept of surface tension as a force created in a surface by cohesive forces had been introduced by von Segner in 1751,[39] but it was not until the middle of the nineteenth century that extensive investigations were undertaken, aimed at measuring the quantity for various liquids and at understanding the factors that controlled the spreading of liquids. It is quite possible that wave-damping, together with the other striking effects produced by oil on water, contributed to the impetus behind such research, as suggested by Giles in 1969.[40] Similarly important were observations of surface movements caused by surface tension, such as the erratic motion of camphor particles on water, of which G. van der Mensbrugghe in 1869 gives a most interesting account.[41]

In 1871, Carlo Marangoni recognized that some of these effects were due to differences in surface tension.[42] Some years before, Plateau had observed that a needle held horizontally in the surface of water and rotated about a central vertical axis showed far more viscous drag than did the same needle held completely submerged in the water. He concluded from this observation that the viscosity of the liquid at the surface must be considerably greater than that of the water at greater depths. Marangoni repeated the experiment, but with the important difference that Plateau's needle was replaced by a horizontal circular ring. If Plateau had been correct in his deduction, then the viscous drag on the rotating ring would have been similarly much greater when held in the surface than when held in the bulk liquid, but Marangoni found that this was not the case. He explained the increased drag in Plateau's experiment in terms of changes in surface tension caused by compression and expansion of the surface, the surface on either side of the rotating needle being either compressed or expanded, and the surface tension being changed so as to give a force opposing the motion. This is, in effect, an elasticity of the surface, a variation of surface stress with surface strain.

Marangoni attempted to use the concept of surface elasticity to explain the damping of waves. His explanation differs considerably from ours and seems to some extent misguided, since he attributes an elasticity to the unoiled surface and not to the oiled surface. However, this apparent confusion may well lie in differences in interpreting the term 'elasticity' itself, and it is reasonable to suppose that his conception of the action of an oil film on waves was essentially the same as that held today:

> Spargendo dell'olio sul mare si sostituisce alla superficie elastica dell'acqua la superficie non elastica dell'olio; sicchè il vento smuoverà localmente le particelle dell'olio senza che il movimento venga a communicarsi per una grande estensione, e di qui il cessare dell'increspamento della superficie coll'effusione dell'olio.

This discovery of the existence of changes in surface elasticity, and of their effects on surface motions, has been of major importance for the development of wave-damping theories, especially for the theory of the damping of waves in the absence of wind. However, it took a long time for the significance of Marangoni's discovery to be widely realized, partly because of the ease with which a clean water surface, with no absorbed surface film and no elasticity becomes contaminated and takes on an appreciable elasticity. We now know that it is only by preparing water with great care, and using it in carefully cleaned apparatus, that we may hope satisfactorily to perform the sort of experiments which were being attempted in the nineteenth century without this knowledge. Another possible reason for the slow acceptance of Marangoni's idea may well be the confusion which has tended to exist between surface tension itself and surface elasticity — the result of variations of the surface tension. The terminological confusion evident in Marangoni's own paper may have been a further factor.

In his fragmentary descriptions of the effects of oil on waves, which will be described in more detail in the next section, Osborne Reynolds shows a more sure grasp of the principles of surface elasticity. Since his publications contain no references at all to earlier work, it is not certain whether he knew of Marangoni's paper, but by 1890 Lord Rayleigh, who was at this time closely concerned with developments in the theory of surface tension as well as with hydrodynamics, credited Marangoni with this crucial discovery.

Before continuing to describe the way the subject of storm-wave damping developed in the last quarter of the nineteenth century, it is interesting to examine the way in which part of the wave-damping problem reached its satisfactory explanation, as a direct result of the discovery of the concept of surface elasticity.

The damping of progressive water waves

Until 1880, the problem of the damping of water waves by oil was considered by many authors to be a single, highly complicated subject. Some observers, such as Captain Müller in 1782, had appreciated that oil had little effect on the swell-progressive gravity waves coming perhaps from storms many hundreds of miles away — and that the major effect of oil was on the immediate tendency of large waves to break under the direct action of the wind. However, others such as Achard, in his wave-tank experiments, appeared to feel that since it was the waves themselves that did the breaking, the method of making them break was not important; they therefore did not consider the presence of wind to be necessary.

In 1825 the brothers Weber had recognized a distinction between the damping of waves in the absence of wind and the effect of oil on the force exerted by the wind on the surface,[32] but this opinion is not evident in the works of Robinet (1807),[30] Frankenheim (1835),[33] or du Bois-Reymond (1858).[34] Marangoni himself does not appear to have considered the effect of elasticity on simple progressive waves, but this may have been because he felt the wind-wave problem to be of more practical interest.

It was Osborne Reynolds who gave the first satisfactory explanation of the effect of an elastic oil film on a water surface. He unfortunately failed to publish his work in a complete form, and we have today only a brief abstract of a paper he gave to the British Association for the Advancement of Science in 1880.[43] The complete abstract is as follows:

> This paper contained a short account of an investigation from which it appeared that the effect of oil on the surface of water to prevent wind-waves and destroy waves already existing, was owing to the surface-tension of the water over which the oil spread varying inversely as the thickness of the oil, thus introducing tangential stiffness into the oil-sheet, which prevented the oil taking up the tangential motion of the water beneath. Several other phenomena were also mentioned. The author hopes shortly to publish a full account of the investigation.

Again, the two aspects of oiling effects are linked, but in this case it is possible to discern from the phrase 'prevented the oil taking up the tangential motion of the water beneath', that Reynolds appreciated that there could be an effect on the damping of the swell itself. We have corroborative evidence from Lord Rayleigh (1890) and from Horace Lamb (1895) to support this conclusion. It was Rayleigh who gave the first clear account of the mechanism involved:

> As the waves advance, the surface of the water has to submit to periodic extensions and contractions. At the crest of the wave the surface is compressed, while at the trough it is extended. As long as the water is pure there is no force to oppose that, and the wave can be propagated without difficulty; but if the surface be contaminated, the contamination strongly resists the alternate stretching and contraction.[44]

Rayleigh acknowledged Marangoni's explanation as 'the first attempt on the right lines', while pointing out the erroneous nature of his explanation.

It was Horace Lamb who published the first mathematical expression of the theory of water-wave damping by an elastic film, in his *Hydrodynamics* textbook of 1895:

> It is evident at once ... that in oscillatory waves any portion of the surface is alternately contracted or extended, according as it is above or below the mean level. The consequent variations in tension produce an alternating drag on the water, with a consequent increase in the rate of dissipation of energy.[45]

Lamb's article contains sufficient analytical results to allow the damping rate to be deduced for any specific value of the elasticity, but he did not perform any detailed calculations, except for the two extreme cases of zero and infinite elasticity. He appears to have assumed that the damping rate increased monotonically from the one extreme value to the other, and the rather short treatment of the problem in the 1895 edition was reduced even further for subsequent editions.

Although V.G. Levich examined the problem in 1940 and 1941,[46] it was not until 1951 that R. Dorrestein, repeating Lamb's calculations, showed that the damping rate passed through a maximum for intermediate values of elasticity and could be very sensitive to small values of elasticity.[47] The subject has since attracted the attention of many scientists, mainly interested in the use of measurements made on systems of high-frequency capillary ripples to deduce the surface-chemical properties of surface-active films. Lucassen-Reynders and Lucassen gave a thorough review of such work in 1969.[1] Extensive series of measurements have now been made for capillary waves, and the agreement of the results with linearized hydrodynamic theory is sufficiently close in the cases of a large number of surface-active films for it to be considered that this aspect of the damping properties of oil films is now well understood.

The second revival of interest in wave-oiling, 1880-1900

Although the results of Marangoni's work eventually allowed Osborne Reynolds and Horace Lamb to lay the foundations of a rigorous theory of the damping of water waves by surface films, his work was not the major cause of the revival of interest in storm-wave damping which took place towards the end of the nineteenth century. We need to look elsewhere to find the reason for the sudden and intense interest in the subject which produced a rush of papers, both scientific and popular, and resulted in the publication of no less than five substantial monographs in an eight-year period from 1887 to 1894. The cause of this vigorous activity was the almost certainly independent rediscovery of wave-oiling by a Scottish manufacturer and businessman, John Shields of Perth (1822-1890). Shields described the origin of his interest in a letter dated 6 May 1882:

> Some five or six years ago, having occasion to be superintending some arrangements in connection with one of the ponds connected with my works here, some oil was spilt on the water, which, at the time, was pretty much broken by a high wind blowing, and to my suprise the oil spread in all directions notwithstanding the wind, which appeared to skim over it, and leave a smooth, glassy surface on the water. This struck me very much, and shortly after, when the pond was in pretty much the same state, I sent for my head mechanic, who got twenty or thirty feet of india-rubber tubing. We then went and laid it along the bottom, keeping one end on land. We then took a flask of oil, and poured it into the tube until it was fully charged; it then began to ascend in beautiful beads to the surface, and spread with lightning rapidity, stilling the whole pond almost instantly, and not more than a gill of oil was used. I was then convinced of the great utility that this could be turned to, and to my mind then, and is still, only a mechanical problem, how to get the oil where it is wanted, and when it is wanted.[48]

Shields performed full-scale experiments in Peterhead harbour, during a hurricane on 1 March 1882, and the attention of the press was drawn to their successful outcome. Coverage in the press soon led to articles in popular journals and magazines, and in a very short time, experiments were being performed all over Europe and in the U.S.A. It is most interesting to observe how Shields' experiments were able to excite such a world-wide revival of interest in a subject which was by no means unknown. Since 1860, *Chambers' Journal* had been publishing accounts of the successful use of the effect, and editorials encouraging its more widespread use.[49] The experiments were discussed in the House of Lords on 28 July 1882, and Shields repeated his tests at Aberdeen harbour on 4 December 1882, in the presence of an observer from the Board of Trade, Captain W. Broughton Pryce.[50]

Although large-scale practical experiments appear to have completely convinced their observers of the value of wave-oiling, they do not appear to have produced any lasting consequences, at least as far as the protection of harbours is concerned. The most immediate and perhaps the most important direct result of the tests was the revival of interest among scientists, notably another Scotsman, John Aitken, and the Belgian, Gustav van der Mensbrugghe. In the same year as Shields' Peterhead and Aberdeen experiments, van der Mensbrugghe wrote a paper in which he described an experimental refutation of Franklin's 'friction-reduction' hypothesis.[51] He found that blowing on the surfaces both of water and of oil gave considerably more ripples than did blowing on an oiled water surface. It was thus demonstrated that the phenomenon could not be due to the wind exerting significantly less friction on the oil molecules than it did on water, and van der Mensbrugghe therefore concluded that the calming effect was associated with the reduction of surface tension by the addition of oil to the surface — a phenomenon of which he had considerable experience. His consequent explanation of wave-oiling, however, leaves much to be desired. He seems to have been confused by the fact that only a very thin layer of oil is needed to remove the greater part of the surface tension of clean water. He appears to have deduced that a thick water layer, consisting of just so many thin water layers added together, would have so many more times potential energy. He reasoned that if the wind acted to slide one thin sheet of water over an adjacent area of clean surface, then the annihilation of the surface energy of the lower surface could appear as kinetic energy of the waves. If all of the water surfaces were covered by layers of oil, however, then he calculated that because an extra interface would result from the same sliding process, there would not just be less surface energy available, there would actually be a deficit of surface energy, and the process would remove kinetic energy from the wave system:

> ... si le vent fait ensuite glisser, en un de ces points, une couche d'eau pure sur une couche recouverte d'huile, le glissement en question, au lieu de produire une dimunution d'énergie potentielle, va remplacer au contraire la surface libre de l'huile par deux autres surfaces ayant ensemble une énergie potentielle bien plus grande que celle de l'huile; de ces deux surfaces, la supérieure est constituée par hypothèse par l'eau non huilée, et l'inférieure est la surface de contact de la couche de l'eau avec la portion de la couche d'huile submergée; l'accroissement d'énergie potentielle ainsi developpé ferait donc naître une résistance toujours croissante, et, en même temps, la tendence de l'huile à remonter à la surface empêcherait le glissement d'une nouvelle couche d'eau. (pp. 13, 14).

Van der Mensbrugghe's concepts of wave motion and of surface tension seem to us rather misguided, but these criticisms cannot be directed at all at the work of the other experimentalist motivated by Shields' results, John Aitken.

Dr. John Aitken, F.R.S. F.R.S.E. (1839-1919), more noted for his contribution to establishing the role of solid particles in the nucleation of rain, reported his laboratory experiments in 1883, the year following Shields' work in Peterhead and Aberdeen.[17] His first experiment, like that of van der Mensbrugghe, was a test of the 'friction-reduction' hypothesis, that the oil film offered less resistance to the passage of the wind than did the clean water surface. He used an air jet, directed obliquely at the surface of water in a circular vessel, to set the water into steady rotation. A submerged paddle suspended from a torsion wire was arranged to measure the torsion exerted by the rotating water. He found that the torsion proved to be little different in the two cases, clean and oiled — perhaps even slightly greater in the oiled case — and the 'friction-reduction' concept was thus shown to be invalid.

Aitken's theory of the damping effect was similar in some respects to that of Marangoni, being based on the resistance to stretching and compression of the film that follows from the film elasticity:

> If we try to move forwards one part of the surface, we find its motion is resisted by the tension of the surface behind it increasing on account of the removal of oil, and the tension in front of the moving area does not increase but diminishes. The forward movement of the film is therefore checked ... The result of this is, that the wind cannot drive forwards isolated patches of film or surface so as to cause waves, but the whole of the surface is caused to advance at nearly a uniform rate, and the formation of waves is thereby prevented. (p. 63).

Aitken saw clearly that if this calming mechanism could operate on a flat surface, then it would also be significant in preventing the generation of small ripples on the surface of large waves, and could be the cause of the reduction of the effect of wind on such waves.

It was at this stage that various government agencies, perhaps pressed by popular interest, began to examine possible uses of the phenomenon in life preservation. One of the first was the United States Life-Saving Service, who investigated the use of oil in near-shore rescue work.[52] Many tests were done on a variety of ingenious devices for spreading the oil, but no satisfactory means was found for maintaining a supply of oil to areas upwind of a distressed vessel or a rescue operation. It was decided that while the phenomenon was doubtless of value in the open sea, especially for vessels running before the wind, its use in surf zones was not practical.

Following experiments by the British Admiralty, the Board of Trade reported recommendations on the best types of oil and the best methods of use,[53] but even so, British Government support was not vigorous. The French made a little more progress towards practical use, following the publication of the work of Admiral Georges-Charles Cloué, a one-time cabinet minister. His monograph had three progressively more informative and more detailed editions in one year, 1887,[54] and a special oil-equipped lifeboat developed to exploit the phenomenon was named after him. Admirals François Paris and Simeon Bourgois also contributed to the debate, in favour of using wave oiling.

Germany and the United States were most vigorous, both in official support of research and in publicizing the results of tests in a form suitable for appreciation by the seamen directly involved in problems of dangerous storms. In the 1880s, the U.S. Hydrographic Office encouraged the reporting of successful instances of wave-oiling, and published details in their widely read monthly *Pilot Charts* series, with year collections being issued separately in 1886 and 1887.[55] In 1887 the Hamburger Nautischen Verein offered a prize for the best book on the subject, and of the twenty-one works submitted (thirteen in English, eight in German) those of E. Rottok and R. Karlowa were awarded prizes and published in 1888.[56] More comprehensive monographs appeared soon, by Josef Grossmann in 1892,[57] and by M.M. Richter in 1894.[58] The oceanographer Wladimir Köppen reported his extensive experiments, using both oil and soap solutions, in his *Annalen der Hydrographie und Maritime Meteorologie* in the 1890s, and encouraged the use of that journal to report both successful instances of oil use and important technical advances.[59]

The explanations offered by Grossmann and Köppen attempted to follow on from the explanation given by Aitken, and were both in terms of a surface elasticity of some sort. Grossmann appears to have confused the roles of surface tension and surface elasticity to some extent, but both authors were principally concerned with possible ways that the surface-tension-reducing property of the oil might modify the wave motion itself, rather than with possible effects of the layer on the stability of the surface against wind disturbances.

The twentieth century — decline of interest

Given the importance ascribed to the subject in the previous twenty years, it is not clear why with the end of the nineteenth century there should have been such a rapid decline in the number of articles being written about wave-oiling, which seems to be associated with a similarly inexplicable decline of interest in the matter. There were still advocates of its more regular and more efficacious use, as can be seen from occasional items published in lifeboat journals and papers given at International Life-Boat Conferences, an article in the 1932 *Proceedings* being particularly instructive in this respect.[60] Perhaps with the increasing application of modern technology to shipbuilding and the increase in size of the more modern vessels, the problems became less important. Another factor may simply have been that the trouble involved in equipping every boat with effective apparatus and training every crew in its use just did not seem justified by the relative infrequency of disasters, bearing in mind improvements in life-saving techniques, and, more recently, the advent of radio communications. However, there have been well attested instances of rescues in the early part of this century in which oil was used with striking results, such as in the wreck of the hospital ship S.S. Rohilla, close to the shore off Whitby on 30 October 1914.[61] In this case, the rescue was finally effected on the second day after the wreck by a lifeboat first spreading oil upwind of the area, and then rushing downwind to take off stranded passengers while the calming effect persisted.

The current picture

Although there is still considerable disagreement among scientists about the importance of the phenomenon, it is becoming more clear how oil might conceivably affect the breaking of dangerous sea waves. It is apparent that oil does little to reduce the energy contained in a highly turbulent ocean-wave system, except on a long time scale, and that its immediate effect is probably on the processes by which energy is continually fed into the system from the wind.

On the open sea in deep water much of the wave breaking appears to be strongly affected, if not directly caused, by the wind acting on individual wave crests. With large-amplitude waves whose size is such that real damage could be caused by their breaking, the turbulent air flow over their crests is likely to separate, and a considerable horizontal force may be applied to a given crest as a result of the aerodynamic drag exerted by the wind on the water surface. This drag will depend on the aerodynamic roughness of the exposed crest, and it is reasonable to suppose that if the oil should act in some way that reduces the surface roughness to a significant extent, then the drag may be sufficiently decreased to give reductions both in the probability of a given wave breaking and in the kinetic energy given to the mass of water thrown off the top of the wave during the breaking event.

So far, this picture barely differs from the earliest discussion of the subject — that of Franklin and also that attributed by Plutarch to Aristotle. It is the next link in the chain of explanation that has confounded generations of scientists, and still lacks a really adequate explanation. How does the oil affect the generation of the short wavelength disturbances on the backs of the large-amplitude waves?

As we have seen, this problem has attracted many more or less implausible solutions over the past two centuries. It is now certain, however, that it is the surface-tension-reducing property of oils that is the crucial factor, rather than their chemical nature, density, or viscosity, that stabilizes the water surface against wave generation by wind. Rather than simply the general decrease in surface tension, it appears likely that it is the mechanical properties of the film that determine the stability — properties such as the dilational elasticity, which follows from the changes in surface tension with compression and extension of the surface.[1] This latter property has been shown to be common to all surface-tension-reducing materials, and has also been found to determine the decay of progressive water waves.

Analysis of several theoretical models involving shear flows near free liquid surfaces has indicated that dilational elasticity generally tends to stabilize the surface against wave excitation. Gottifredi and Jameson recently showed that if an elastic boundary condition is incorporated into the model proposed by Miles for the generation of water waves by a laminar shear flow in the adjacent air, the stability so produced has many of the characteristics common to the effects of oil on sea waves.[62] It was predicted that, as is observed, the shorter wavelengths would be the most reduced by the surface film. While this indicates that the elasticity is very probably of importance in the practical case however, it still does not completely solve the problem. The Miles model is highly idealized, and its relevance to the clean water case has been seriously questioned.

Thus we are still lacking a completely satisfactory theory of wave-oiling. Until one is found, and until there has been a really convincing experimental demonstration that wave-oiling does produce significant calming and a reduction of danger, the subject is likely to remain in the same state of optimistic uncertainty as has marked its progress so far.

Notes

I am grateful to the Natural Environment Research Council for the provision of my fellowship at the University of Essex. Many people have kindly assisted with the collection of the information contained in this paper, including Mr. T. Tostevin and the interlibrary loans staff of the University of Essex; Mrs. V. Riley of the Naval Historical Library; Mrs. S. Doran, translator, of Norton-on-Tees; Hr. V. Groothoff of the Royal Netherlands Academy of Sciences; Mr. R. Kipling of the Royal National Lifeboat Institution; and Mme. L. Nicholas of the Académie Royale de Belgique.

1. The surface dilational elasticity of a liquid surface is defined as the proportional change of surface tension caused by an infinitesimal change of the finite area of the surface under consideration:

$$\epsilon = \frac{\delta\gamma}{\dfrac{\delta A}{A}} \equiv \frac{\delta\gamma}{\delta(\ln A)}$$

Its units are the same as those of surface tension. In practice, the elasticity depends on the time scale of the surface-area changes involved, unless the surface tension is determined by the concentration at the surface of a fixed quantity of insoluble surface-active material. If the material is soluble, then the surface tension is determined by an equilibrium between surface- and bulk-concentrations — an equilibrium which needs time for it to be achieved. Pure liquids do not exhibit surface dilational elasticity, except perhaps at time scales considerably less than 1 μs. An excellent account of the properties of elastic surfaces, specifically relevant to their effects on wave propagation was given by E.H. Lucassen-Reynders and J. Lucassen, 'Properties of Capillary Waves', *Advances in Colloid and Interface Science*, 2, 347-95, 1969.

2. Plutarch, *Moralia*, Loeb Classical library, Vol. 11, 'Quaestiones Naturales', No. 12, translated by F.H. Sandbach, Heinemann, 1965.

3. Aristotle, *Problemata*, Book 23, No. 38, translated by W.S. Hett, Heinemann, 1937. It is widely accepted that Aristotle himself was not the author of much of the *Problemata*.

4. Plutarch, *Moralia*, Loeb Classical library, Vol. 6, 'De Primo Frigido', No. 950, translated by W.C. Helmbold, Heinemann, 1939 (repr. 1967).

5. C. Plinius Secundus, *Historia Naturalis*, Book 2, Ch. 103. This work was completed in A.D. 77, and of the many editions and translations published since then, this item appears variously as Ch. 106. It begins 'omne oleo tranquillari'. There is an element of confusion evident in references to Pliny on this subject, as some authors have attributed to him the following passage: 'Ea natura est olei, ut lucem adferat et tranquillet omnia, etiam mare, quo non aliud elementum est implacabilis'. This passage originates as a footnote appended to the 1693 Amsterdam edition of Erasmus' 'Naufragium' (see note 11) [British Museum 57.m.14].

6. W.R. Barger, W.D. Garrett, E.L. Mollo-Christensen, and K.W. Ruggles, 'Effects of an artificial slick upon the atmosphere and the ocean', *Journal of Applied Meteorology*, Vol. 9, pp. 396-400, 1970.

7. *Dialogue de Théophylacte Simocrate sur diverses questions naturelles utiles et plaisantes et leur solutions*; translated from Greek into French by Féderic Morel, Paris, 1603. The relevant passage is given in *Revue des Cours Scientifique de la France et de l'Étranger*, Vol. 33, p. 192, 1884, and in an Italian translation by E. Mancini, in *Nuova Antologia di Scienze, Lettere, ed Arti*, Vol. 78, pp. 273-87, 1884.

8. The Venerable Bede, *Historia Ecclesiastica Brittanorum*, Book 3, Ch. 15, translated by B. Colgrave and R.A.B. Mynors, *Bede's Ecclesiastical History of the English People*, Oxford University Press, 1969, pp. 260-1.

9. H. Canisius, *Antiquae Lectionis*, Ingolstadt edition, 1604, p. 696.

10. S. Majolus, *Dies Caniculares*. (Hoc est colloquia tria et viginti physica . . .) 1607, pp. 384-5.

11. [Desiderius] Erasmus Roterodamus, *Colloquia*: 'Naufragium', Simonem Colineum, Paris, 1527, p. 161.

12. B. Franklin, *Philosophical Transactions of the Royal Society of London*, Vol. 64, pp. 445-60, 1774. This paper has been reprinted and translated many times. Frans van Lelyveld published Dutch and French versions in 1775 (note 13), and popular abridgements also appeared at this time, e.g. *Gentleman's Magazine*, February 1774.

13. F. van Lelyveld, *Berichten en Prijsvragen, over het storten van Olie, Traan, Teer, of andere dryvende Stoffen, in Zee-gevaren*, Leiden, 1775; *Mémoire sur l'usage des huiles, du goudron, et de toute matière qui surnage, pour diminuer les dangers sur mer*, Amsterdam, 1775; *Essai sur les moyens de diminuer les dangers de la mer par l'effusion de l'huile, du goudron, ou de toute autre matière flottante*, Amsterdam, 1776.

14. M. Wall, 'Some observations on the phaenomena, which take place between oil and water', *Memoirs of the Literary and Philosophical Society of Manchester*, Vol. 2, pp. 419-28, 1785.

15. Apart from the references made by Franklin, there had been other roughly contemporary references: by Martin Martin, in his *Description of the Western Islands of Scotland*, London, 1703, p. 48, (reprinted in *Gentleman's Magazine*, March 1775); by Vitaliano Donati, who described his technique of stilling water surfaces to allow examination of underwater life in his *Della Storia Naturale Dell'Adriatico*, Venice, 1750, pp. 16-17; and by Louis Poinsinet de Sivry, who called for modern verification of the effect in his French translation of Pliny, published in 1771.

16. T. Pennant, 'Seal', *British Zoology*, Vol. 4, No. 46, London, 1770.

17. J. Aitken, 'On the effect of oil on a stormy sea', *Proceedings of the Royal Society of Edinburgh*, Vol. 12, pp. 56-75, 1883.

18. R.N. Ivanov, 'The cause of the damping of waves by means of oil', *Izvestiya Akademii Nauk SSSR, Seriya Geofizicheskaya*, No. 3, pp. 325-43, 1937 (in Russian with English abstract).

19. J.G.D. Müller, 'Ueber den Gebrauch, den Seeleute im Sturm vom Oehl machen, das Brechen der Wellen zu verhindern', *Göttingisches Magazin der Wissenschaften und Litteratur*, Vol. 2, No. 6, pp. 323-41, 1782.

20. Mann, 'Mémoire dans lequel on examine les effets et les phénomènes produits en versant différantes sortes d'huiles sur les eaux, tant tranquilles qu'en mouvement, d'après une suite d'experiences faites à ce sujet', *Memoires de l'Academie Imperiale et Royale des Sciences et Belles-Lettres de Bruxelles*, Vol. 2, pp. 257-94, 1780.

21. P. Frisi, 'Dell'azione dell'olio sull'acqua', *Dissertazione III, Opuscoli Filosofici*, Milano, pp. 49-66, 1781.

22. A.L.F. Meister, 'De quibusdam olei aquae superfusi effectibus opticis et mechanicis', *Commentationes Societatis Regiae Gottingensis*, Vol. 1, Sec. 2, pp. 35-64, 1778.

23. I have not yet traced this work, quoted by van Beek in 1842, who gives the reference as: 'Ernstige berispingen en aanmerkingen op de berigten en prijsvragen over het storten ...' (see note 13) 'door den Heer Frans van Lelyveld, rondborstig en Vaderlandlievend opgegeven, ter handhaving van de eere der nederlandsche en bijzonder der Zeeuwsche visschers en zeelieden in die voorvaderlijke Kundigheid, door Joannes Le Francq van Berkhey, enz., te Leyden bij Heyligert en Hoogstraten, 1775'.

24. W.J. Millar 'Why oil has an influence on waves', *The Shipping World*, p. 176, 1 October 1893.

25. It seems that Achard's work first appeared, in a French version, in the *Nouveaux Mémoires de l'Académie Royale des Sciences et Belles-Lettres de Berlin*, for the year 1778, 'Mémoire sur la manière de calmer l'agitation d'une partie de la surface d'un fluide, soit par l'affusion d'un fluide spécifiquement léger ...', pp. 19-26, published in 1780. For some reason, this article was not referred to by contemporary authors, even though there were references (e.g., Müller, 1782[19]) to the relevant experiments considerably before the usually cited account in German in Achard's own collection of 1784, 'Ueber die Art, die Wellenförmige Bewegung eines Theils der Oberfläche einer Flussigkeit zu vermindern ...', in *Sammlung physikalischer und chemischer Abhandlung*, Berlin, Vol. 1, pp. 83-93, 1784.

26. A. van Beek, 'Over het golvenstillend vermogen der olie ...', Utrecht, 1841, translated into French as 'Mémoir concernant la propriété des huiles de calmer le flots, et de rendre la surface de l'eau parfaitement transparente', *Annales de Chimie et de Physique*, 3me Serie, Vol. 4, pp. 257-89, 1842, and from thence into German in *Annalen der Physik* (Poggendorff), 57, 419-52, 1842.

27. J.F.W. Otto, 'Das Oel ein Mittel die Wogen des Meeres zu besänftigen', *Allgemeine Geographischen Ephemeriden*, Vol. 2, pp. 516-27, 1798.

28. F.C. Kries, 'Einige Bermerkungen über J.F.W. Otto's Aufsatz: Das Oel ...', *Allgemeine Geographischen Ephemeriden*, Vol. 3, pp. 242-51, 1799.

29. F. Gerstner, 'Theorie der Wellen', reprinted in *Annalen der Physik*, Vol. 32, pp. 412-45, 1809.

30. J. Robinet, 'Explication d'un phénomène d'hydrostatique observé par Franklin', *Journal de Physique, de Chimie, et d'Histoire Naturelle*, pp. 277-82, 1807.

31. F.X. de Zach, *Correspondance Astronomique, Geographique, Hydrographique, et Statistique*, Vol. 6, pp. 489-512, 1822.

32. E.H. Weber and W. Weber, 'Ueber die Besänftigung der unter dem Einflusse des Windes erregten Wellen durch die Ausbreitung von Oelen auf der Oberfläche von Wasser', *Wellenlehre, auf Experimente gegründet*, Leipzig, 1825, pp. 60-90.

33. M.L. Frankenheim, 'Beruhigung der Wellen', *Die Lehre von der Cohäsion*, Breslau, 1835, pp. 148-52.

34. P. du Bois-Reymond, 'Experimentaluntersuchung über die Erscheinungen, welche, die Ausbreitung von Flüssigkeiten hervorruft', *Annalen der Physik und Chemie* (Poggendorff), Vol. 104, pp. 193-234, 1858.

35. P. van Griethuyzen, *Jets of niets, of invallende gedachten over mogelijke voorbehoedmiddelen, ter beveiliging onzer zeedijken en zeeweringen tegen zware zeestortingen en golfslagen*, Utrecht, 1834.

36. P. de Leeuw, *Algemeene Konst-en Letterbode*, No. 10, pp. 157-9, 1837.

37. 'Verslag eener Commissie ... over het golvenstillend vermogen der olie', Het Instituut (Royal Netherlands Institute) 1841, No. 2, pp. 89-91. This report was transcribed by van Beek in 1842 (note 38). The gist of the report was published in Paris two years later, 'Note sur les expériences faites par une Commission ... dans le but de vérifier la propriété

attribuée à l'huile de calmer les vagues de la mer', *Comptes Rendus Hebdomadaires de l'Académie des Sciences*, Paris, Vol. 16, pp. 399-401, 1843. Lipkens added his dissenting contribution in the same volume, pp. 684-6.

38. A. van Beek, *Aanmerkingen op eene proef over het golvenstillend vermogen der olie*, Utrecht, 1842.

39. A. von Segner, 'De figuris superficierum fluidarum', *Commentarii Societatis Regiae Scientiarum Gottingensis*, Vol. 1, pp. 301-36, 1751.

40. C.H. Giles, 'Franklin's teaspoon of oil. Studies in the early history of surface chemistry Part I', *Chemistry and Industry*, pp. 1616-24, 1969.

41. G. van der Mensbrugghe, 'Sur la tension superficielle des liquides, considerée au point de vue de certains mouvements observés à leur surface', *Mémoires couronnés et mémoires des savants étrangers, publiés par l'Académie Royale des Sciences, des Lettres, et des Beaux-Arts de Belgique*, Vol. 34, 67pp., 1869.

42. C. Marangoni, 'Sul principio della viscosità superficiale dei liquidi stabilito dal sig. J. Plateau', *Nuovo Cimento*, Series 2, Vol. 5/6, pp. 239-73, 1872.

43. O. Reynolds, *Report of the British Association for the Advancement of Science*, 1880, pp. 489-90.

44. Rayleigh, *Proceedings of the Royal Institution*, Vol. 13, pp. 85-97, 1890.

45. H. Lamb, 'The calming effect of oil on water waves', *Hydrodynamics*, 2nd ed., Cambridge University Press, 1895, Art. 304, pp. 552-5.

46. V.G. Levich, 'The damping of waves by surface-active materials', *Zhurnal Eksperimentalnoi i Teoreticheskoi Fiziki*, Vol. 10, No. 11, pp. 1296-1304, 1940; also *Acta Physicochimica URSS*, Vol. 14, pp. 307-28, 1941.

47. R. Dorrestein, 'General linearized theory of the effect of surface films on water ripples', *Proceedings, Koninklijke Nederlandse Akademie van Wetenschappen*, Series B, Vol. 54, pp. 260-72, 1951.

48. Given as an appendix to the report by B.C. Sparrow cited in note 52.

49. *Chambers' Journal*, pp. 25-7, pt. 2, 1860; pp. 511-12, 1878; pp. 811-14, 1878; pp. 497-8, 1879; pp. 140-2, 1883.

50. Pryce's report is appended to that of B.C. Sparrow cited in note 52. Further details of Shields' experiments are given by C.G. Gordon Cumming, 'Oiling the waves — a safeguard in tempest', *The Nineteenth Century*, Vol. 11, pp. 572-85, 1882 (also included in B.C. Sparrow cited in note 52); and by C.H. Giles and S.D. Forrester, 'Wave damping — the Scottish contribution', *Chemistry and Industry*, pp. 80-87, 1970.

51. G. van der Mensbrugghe, 'Sur les moyens proposés pour calmer les vagues de la mer', *Bulletins de l'Académie royale de Belgique*, Ser. 3, Vol. 4, No. 8, 1882; also published separately, 23pp., in Brussels the same year.

52. B.C. Sparrow, 'The use of oil in calming rough seas', *Annual Report of the United States Life-Saving Service*, pp. 427-87, 1883.

53. 'Official report on the use of oil at sea, for modifying the effect of breaking waves', *Board of Trade Journal*, Vol. 1, No. 4, pp. 211-12, 1886.

54. G.-C. Cloué, 'Le filage de l'huile': first edition in *Annales Hydrographiques*, Vol. 2, 1887; second edition, 40pp., Imprimerie Nationale, Paris, 1887; and third edition, 105pp., Gauthier-Villars, Paris, 1887.

55. G.L. Dyer 'The use of oil to lessen the dangerous effect of heavy seas', *U.S. Hydrographic Office Report* No. 82, Washington D.C., 27pp., 1886; E.B. Underwood, U.S. Hydrographic Office Report No. 83, 22pp., 1887.

56. H. Seemann 'Die Beruhigung der Sturz-Wellen durch Anwendung von Öl', *Das Wetter*, 5, 239-49, 1888; E. Rottok, *Die Beruhigung der Wellen durch Ol*, Berlin, 1888; R. Karlowa, *Die Verwendung von Oel zur Beruhigung der Wellen*, Hamburg, 1888.

57. J. Grossmann, *Die Bekämpfung der Sturzwellen durch Öl und ihre Bedeutung für die Schiffahrt*, Vienna, 1892.

58. M.M. Richter, *Die Lehre von der Wellenberuhigung*, Berlin, 1894.

59. There are about 39 separate items concerning wave-oiling in Vol. 15-25, 1887-97. Köppen's own significant contribution, 'Verhalten der Oele und Seifen auf Wasseroberflächen und Rolle der Oberflächenspannung bei Beruhigung der Wellen', appeared in pp. 134-49 of Vol. 21, 1893.

60. J.D. Banting, (with an introduction by H. Th. de Booy) 'The influence of oil on water', *Proceedings of the 3rd International Lifeboat Conference*, Rotterdam, Holland, pp. 155-73, 1932.

61. 'The wreck of the S.S. Rohilla', *The Life-Boat*, Vol. 22, pp. 198-203, 1915.

62. J.C. Gottifredi and G.J. Jameson, 'The suppression of wind-generated waves by a surface film', *Journal of Fluid Mechanics*, Vol. 32, pp. 609-18, 1968.

The Contributors

JOYCE BROWN is a Lecturer in the Environmental Studies section of the Civil Engineering Department at Imperial College, London and has specialized for a number of years in biographical and historical studies.

R.A. BUCHANAN is Director of the Centre for the History of Technology, University of Bath, and author of numerous works on the history of technology and industrial archaeology.

A. RUPERT HALL is Professor of the History of Science and Technology at Imperial College, London.

P.B. MORICE is Professor of Civil Engineering at the University of Southampton.

FRANK D. PRAGER was a U.S. patent attorney who had written studies of the lives and works of Brunelleschi and Taccola.

JAMES A. RUFFNER has for many years taught history of science and technology at Wayne State University, Detroit, Michigan, U.S.A.

JOHN C. SCOTT is a Fellow of the Fluid Mechanics Research Institute at the University of Essex.

JACK SIMMONS is Emeritus Professor of History, University of Leicester, and founder and former editor of the *Journal of Transport History*.

T
14.7
H57
1978

AUG 1 1979